Applied Cartography
Source Materials for Mapmaking

Thomas D. Rabenhorst

University of Maryland
Baltimore County

Paul D. McDermott

Montgomery College

Merrill Publishing Company
A Bell & Howell Information Company
Columbus   Toronto   London   Melbourne

Published by Merrill Publishing Company
A Bell & Howell Information Company
Columbus, Ohio 43216

This book was set in Palatino

Administrative Editor: David Gordon
Developmental Editor: Wendy Jones
Production Coordinator: Rex Davidson
Art Coordinator: Ruth Kimpel
Cover Designer: Cathy Watterson

Library of Congress Catalog Card Number: 88–61054
International Standard Book Number: 0–675–20533–6
Printed in the United States of America
1 2 3 4 5 6 7 8 9—92 91 90 89

To Our Wives—Gayl and Carol

PREFACE

Applied Cartography: Source Materials for Mapmaking is the first book in a series on the creation of maps. Each of the books focuses on a particular facet of map construction. Each includes specifically designed student projects aimed at reinforcing the cartographic concepts presented. This collection provides instructors with a variety of books to choose from to best meet the needs of their individual cartography courses.

The series is unique in several respects. In the first place, a balance between theory and practice has been achieved. The books discuss cartographic concepts and then reinforces them with specifically designed projects for practical application. Many projects are included that simulate tasks frequently encountered in the workplace. Second, the variety of cartographic projects in the series enables instructors to focus on those best suited to meeting their course requirements. Because many departments operate with limited budgets, the authors have attempted to provide all the data, base maps, and other mapping source materials as they could for a given project. While the length of project completion time is left to the instructor's discretion, each project has been designed so that it can be accomplished in one to two weeks. Third, "cameos" written by cartographic experts demonstrate in a most interesting way the variety of work undertaken by cartographers. These cameos amplify some of the concepts introduced in each section and provide the reader with an added dimension to the subject being discussed.

This first volume in the series addresses the issue of source materials used in compiling maps. Before a map can be compiled, cartographers must collect all the information necessary for creating the map. They must decide on an appropriate base map or maps, locate or collect specific data, sometimes utilize aerial photographs or other forms of imagery, and occasionally go into the field to actually create a base map or locate necessary features to be included in the map. Frequently, cartographers spend a great deal of time investigating and obtaining source materials before the actual task of map construction begins.

Applied Cartography: Source Materials for Mapmaking provides students with an overview of available source material and suggests specific directions to follow in gathering this material. Section 1 concentrates on maps as source materials. It addresses the task of locating previously published maps as well as aerial and satellite imagery. Often, much unnecessary work can be avoided by a thorough search of existing maps; this section advises students where to look. At the end of the section a series of exercises is included to refine the student's investigative skills. Ronald E. Grim, head of the Reference and Bibliography Section at the Library of Congress, provides an informative cameo with an excellent overview of the many sources available to the cartographer and researcher alike.

Section 2 reviews some basic methods of field mapping. Occasionally, cartographers must actually go into the field to create a base map or collect map-related data. This happens most often in mapping on a very large scale. Detailed coverage is given to foot-made maps, compass traverses, and simple plane-table mapping used to map relatively small areas at large scales.

Section 3 examines the ever-valuable topographic map as the most readily available and frequently most accurate of source maps. After pointing out the various topographic series available, discussion centers on the wealth of information contained in the topographic map. Both the marginal information as well as the mapped data are examined. As in the other sections, a series of exercises is included to help the student apply the concepts presented in the text. Morris M. Thompson, author of *Maps for America*, produced by the USGS, provides an outstanding overview of the topographic mapping process in his cameo presentation.

The intent of this text is to provide a "broad-brush" approach to pursuing source materials for map construction. Rather than singling out any one topic for particular emphasis, direction and guidance on a broad range of topics are given. In this way the student will begin to grasp the importance of available source materials and the processes by which to obtain them.

The authors wish to thank the following reviewers of the manuscript for their helpful comments and suggestions: Barbara Buttenfield, University of Wisconsin; Michael DeMers, Ohio State University; James F. Fryman, University of Northern Iowa; Roland Grant, Billings, Montana; Janet Gritzner, South Dakota State University; Alan MacEachren, Pennsylvania State University; Gregg Plumb, University of Oklahoma; and Richard D. Wright, San Diego State University.

INTRODUCTION TO THE SERIES

Applied Cartography: Source Materials for Mapmaking is the first in a series of books, each of which emphasizes a particular aspect in the production of maps. Each book can be used independently of the others. Consequently, each contains all the essential information and illustrative material necessary for acquiring an introductory understanding of a particular stage of map construction. Instructors can select from the volumes those that best fit their individual needs.

An additional feature of these books are the "cameo" presentations. In the cameos, recognized authorities from the field of cartography express their unique viewpoints and relate their own experiences relevant to the topic of the text. In a personal way, cameos expose the reader to a variety of insights.

Of particular interest is the project focus of the series. In order to complement the concepts and mapping techniques, the authors offer practical, applied projects through which students gain experience in the actual production of maps. Instructors can elect to use all or some of the projects. All of them can be completed in a reasonably short period of time. Furthermore, since they all include most of the graphics and forms necessary for their completion, department or individual expenditures are reduced. In some cases, special forms have been designed to facilitate the organization of data needed for a project. A list of materials and equipment is presented for each one, along with a sequence of steps (procedures) to be followed. This highly structured approach is particularly useful for the inexperienced student.

It has been noted that in many of the cartography texts currently available the creation or production of large-scale maps is not given sufficient attention. Yet, there is a great opportunity for employment for graduate geographers/cartographers in such agencies as the Defense Mapping Agency and the United States Geological Survey where large-scale maps are a major production concern. It should be noted, too, that the geographer/cartographer can sometimes comprehend the construction problems of small-scale maps by examining some of the procedures used to construct large-scale maps and the manner by which content on a larger scale is modified to create a smaller-scale product. Hence, whenever practical, the authors have included material aimed at simulating part of the large-scale mapping process and the types of problems encountered by professionals working in government or large, private mapping agencies.

The authors of the books in this series encourage your comments and suggestions for improvements. Ideas for other innovative projects or additional topics to be covered, which would complement or replace those in the current editions, would be welcomed. Please send your comments or suggestions to either Professor Paul D. McDermott, Department of Applied Technologies, Montgomery College, Rockville, Maryland 20850 or to Thomas D. Rabenhorst, Director of Instructional Cartography, Department of Geography, University of Maryland-Baltimore County, Catonsville, Maryland 21128.

CONTENTS

Preface   v

Introduction to the Series   vii

Introduction   1

1   SOURCE MATERIALS FOR MAPPING   3

Creating New Maps from Existing Ones   3

Locating Previously Published Maps   5

Defining the Source-Map Parameters   5

Location / Scale / Timeliness of Source Material /
Subject Material

National Sources for Maps and Imagery   7

Library of Congress / National Archives / National
Cartographic Information Center (NCIC) / Satellite
Imagery / Index Sheets

Obtaining Aerial Photographs   10

Aerial Photography Summary Record System (APSRS) /
Private Sources of Aerial Photography

Satellite Imagery   15

Annotation of LANDSAT Imagery / Acquisition of
Satellite Imagery / Ordering LANDSAT and SPOT
Imagery

Cameo: Dr. Ronald E. Grim 22

Introduction / How to Find Important Map Collections / The Library of Congress Map Collection / National Archives Collection / Cartobibliographies / Locating Base Maps

Selected Readings 31

Sources of Maps and Remote-Sensing Imagery 32

Project 1A: Identifying Source Maps 35

Project 1B: Cataloging a Map 43

Project 1C: Determining Availability of Aerial Photography 49

Project 1D: Image Location: Path and Row Indexing 51

Project 1E: LANDSAT Image Annotation 55

2 FIELD MAPPING 57

Introduction 57

Foot-Made Maps 57

Compass-Traverse Maps 58

Closed Compass Traverse / Closing the Traverse / Filling in the Detail / Open Compass Traverse

Plane-Table Maps 62

Selected Readings 63

Project 2A: Foot-Made Maps 65

Project 2B: Reconstructing a Foot-Made Map from Field Notes 71

Project 2C: Reconstructing and Closing a Compass Traverse from a Field Sketch 73

Project 2D: Collecting Field Data for a Closed Compass Traverse 75

Project 2E: Plane-Table Mapping 79

3 TOPOGRAPHIC MAPS 83

Introduction 83

Map Scale 83

Quadrangle System of Map Layout 84

Map Grids 85

United States Public Lands Survey / Universal Transverse Mercator (UTM) Grid / State Plane Coordinate Grid

Series Maps and Special Maps 94

Selection of Mappable Features 95

Map Symbols, Colors, and Labels   98

Relief Information   98

Contour Interpolation / Profiles

Information Shown on Map Margins   103

Map Identification / Lower Margin Data / Projection and Grid Labels / Other Marginal Data

Map Revision   108

Total Revision / Partial Revision / Photorevision / Photoinspection

Cameo: Morris M. Thompson   112

Introduction / Elements of a Topographic Map / Topographic Mapping Operations / New Trends in Topographic Mapping

Selected Readings   115

Project 3A: Topographic Map Interpretation   117

Project 3B: Cross-Sectional Profiles from Topographic Maps   121

Project 3C: Landscape Visualization from Topographic Maps   125

Project 3D: Map Revision   127

Project 3E: Contour Interpolation   133

Introduction

The construction of a map begins when someone needs to illustrate a spatial relationship at a specific place. At this point, there are two possibilities: (1) either a map already exists, and must be sought, or (2) a map does not exist, and must be created. Since it is seldom desirable to "reinvent the wheel," you, as the cartographer, should first conduct a cartobibliographic search to determine the existence of such a map. This is easier said than done, since the resources available for this kind of search vary widely.

If a reasonable search has been conducted, and a satisfactory map cannot be found in a form that will fulfill the need, you must create the map. Through proper data collection and access to existing graphics, you can begin map construction. Your initial problem will be to locate and obtain the appropriate "base" map material.

In most cases, you will find the necessary data on maps produced and published by a wide variety of agencies. Your task as cartographer is to identify the appropriate agency and the graphics essential to the job. Usually, the key products are referenced on index maps, and the specific items may be found in library collections, archives, or map depositories. Frequently, you must purchase these products or have copies made from the originals.

Complementing maps as sources of data are aerial photographs and other remotely sensed imagery. This imagery frequently is the primary source of data for large-scale maps, and often is used to update individual map features. Increasingly, smaller-scale land-use mapping has been aided by the use of satellite imagery.

When neither existing maps nor imagery provide the data necessary to begin map construction, the alternative is to conduct a field survey. Field surveys provide personal contact with the study area. In a field survey, you conduct observations and measurements to obtain the required data. You record data in notebooks, along with photographs and sketches made at the site. Later, this material is used in constructing the map.

Although maps, remote-sensing imagery, and field work provide the essential data or base materials for the map, other data sources are available. You should be aware of their wide variety. As a cartographer, you may resort to books, census reports, wills, inventories, diaries, drawings, or data that has been gathered through interviews and questionnaires. Data is often derived using instruments, such as surveying transits, stream gauges, and weather instruments, to list a few. During a typical career in mapmaking, you will exploit a number of these sources.

In this book, it is not possible to discuss all of the materials which could be used in the compilation or construction of maps. Instead, this text overviews key sources of map-compilation data, in three sections: (1) *Source Materials for Mapping*, (2) *Field Mapping*, and (3) *Topographic Maps*.

The first section, *Source Materials for Mapping*, guides you in locating the maps, photographs, and imagery used in map compilation (base map construction). The necessity for a study in sources of map materials became obvious to Professor McDermott in a "Map Intelligence" course presented by Professor John Sherman at the University of Washington. Other ideas for this section came from the training requirements for employees in libraries that maintain map collections. The text presents a variety of projects duplicating actual map-production tasks performed by cartographers.

The second section, *Field Mapping*, has a different objective. Elementary field-mapping techniques that can be performed by the novice cartographer are presented. The section provides exposure to the procedures and problems of collecting data in the field.

It has been our experience that creative and fruitful field work (to be used as student projects) is best accomplished in interesting settings, such as state or national parks, historical sites, or ecological preserves. An additional benefit is the support frequently given by park personnel in the form of facilities, equipment, and information to the students and teachers.

The third section features *Topographic Maps*. As source materials, topographic maps are indispensable to the cartographer. From them a variety of map products have evolved, including smaller-scale topographic maps, bases for thematic maps, and terrain models. Further, the quality and content of products developed from topographic maps is controlled by the skill used to extract the data contained within the topographic map. Therefore, the objective of this section is to describe the contents of this keystone product, and to provide you with experience in the extraction of information from this kind of map.

1

Source Materials for Mapping

CREATING NEW MAPS FROM EXISTING ONES

While information used in the construction of maps is derived from a variety of sources such as tabular data, historical records, and verbal accounts, previously published maps comprise a major source of information. In government agencies, large-scale maps often serve as the information base from which smaller-scale maps are constructed. For example, a topographic quadrangle map with the scale of 1:24,000 is used by the United States Geological Survey (USGS) as the primary information source for maps drawn at the scale of 1:250,000. In the case of the Defense Mapping Agency, 1:50,000-scale maps are used to create 1:250,000-scale maps. In both situations, a number of larger-scale products are assembled and reduced in size to create a single smaller-scale product. To create a single small-scale map at 1:250,000 from 1:24,000-scale maps requires the extraction of information from about 128 quadrangles (figure 1.1). Using 1:50,000-scale maps as the base requires approximately 25 maps.

In addition to requiring a large number of maps to create a small-scale product, the construction process theoretically requires the cartographer to eliminate about 90% of the detail found on the 1:24,000-scale series when compiling at the 1:250,000 scale. Roughly 80% of the detail on the 1:50,000-scale series will be lost in a similar compilation process. Obviously, losses are reduced if the difference in scale between the source and final map is lessened. The amount of information eliminated can be visualized by reducing the size of a 7.5-minute, 1:24,000 quadrangle to fit its areal location on the 1 × 2-degree, 1:250,000 base. The 1:24,000 map will represent about 6% of the linear dimension, and occupy less than 1% of the area (figure 1.1).

Thematic or special-subject maps also may be derived from large-scale maps to create a smaller-scale graphic (figure 1.2). In general, thematic maps feature fewer variables than are present on topographic or reference maps. Frequently, thematic maps are less concerned with precise location of features; instead, they emphasize the distribution of the subject over an

FIGURE 1.1
U.S. Geological Survey map series coverage. Areal coverage of five different map series is shown. The gray area in each rectangle illustrates the amount of ground portrayed on a single map of that series. As the map scale becomes larger (going from 1:250,000 scale toward 1:24,000), a greater number of sheets is required to provide map coverage of the area. (USGS)

USGS MAP SERIES COVERAGE

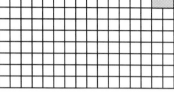

7.5 - MINUTE SERIES
Map Scale *. 1:24,000
Map to ground ratio . . . 1 inch represents 2000 feet
Area covered . . . . 49-70 square miles
Paper size (approx.) . . . . 22 inches by 27 inches
Contour and elevations ** . . . . shown in feet

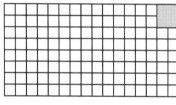

7.5 x 15 - MINUTE SERIES
Map Scale 1:25,000
Map to ground ratio . . . .1 inch represents 2083 feet
Area covered . . . . 100-140 square miles
Paper size (approx.) . . . . 24 inches by 40 inches
Contour and elevations . . . . shown in meters

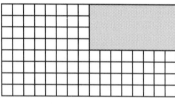

15 - MINUTE SERIES
Map Scale *** 1:62,500
Map to ground ratio . .1 inch represents about 1 mile
Area covered . . . . 197-282 square miles
Paper size (approx.) . . . . 18 inches by 22 inches
Contour and elevations ** . . . . shown in feet

30 x 60 - MINUTE SERIES
Map Scale 1:100,000
Map to ground ratio . . . . 1 inch represents 1.6 miles
Area covered . . . . 1578-2167 square miles
Paper size (approx.) . . . . 29 inches by 44 inches
Contour and elevations . . . . shown in meters

1 X 2 - DEGREE SERIES
Map Scale 1:250,000
Map to ground . . . . .1 inch represents about 4 miles
Area covered . . . . 4580-2167 square miles
Paper size (approx.) . . . . 22 inches by 32 inches
Contour and elevations ** . . . . shown in feet

* 1:25,000 scale on selected maps
** Shown in meters on selected maps
*** 1:50,000 scale on selected maps

area or region. Consequently, thematic maps tend to be drawn at smaller scales than topographic maps.

Oftentimes, the thematic map is the result of information from several data sources, including tabular or statistical materials, verbal descriptions, assorted maps, and field surveys. The compilation process is often compounded by the difficulty of fitting data collected from these varied sources into a common base. To create a special-subject map, you, the cartographer, must make a number of decisions, including the appropriate base map, type of projection, scale of the base map, and the kinds of information necessary to create the final map. Frequently it is the cartographer who must decide how to arrange and locate this information on the new map.

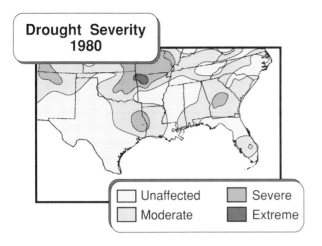

FIGURE 1.2

Thematic map. A thematic map displays the location and distribution of a specified subject. This map illustrates drought conditions in the southern United States. Each shaded area represents a different level of drought condition, described in the legend using the terms *severe, moderate,* or *extreme.* The type of symbolization employed is isopleth.

LOCATING PREVIOUSLY PUBLISHED MAPS

Previously published maps are primary sources of information for new maps. So, in this section, we emphasize finding and retrieving existing maps to be used for the compilation of new maps. We will discuss major libraries that maintain significant collections, the use of index maps to determine the availability of source materials, and the creation of a master index to demonstrate the availability of information.

Maps are produced in prodigious quantities by a variety of agencies throughout the world. Your initial problem when beginning the compilation and construction of a new map is to find out what has been produced, and by whom. As Phillip Muehrcke notes in his book, *Map Use: Reading, Analysis, and Interpretation,* "In view of these vast cartographic resources, it is somewhat paradoxical that locating appropriate maps is one of the map user's greatest obstacles" (p. 359). Once the potential source maps have been located, you must evaluate these maps for possible use.

DEFINING THE SOURCE-MAP PARAMETERS

Location

Your search for source maps must follow a logical path. The first step is to define the location (region/area) to be represented on the new map. Location can be defined in two ways. One is the *political* definition: is the area an entire country, state, county, province, or municipal region? The other defining parameter is the *longitude and latitude* of the region. Longitude and latitude are terms that define angular measurements on spheres such as the Earth. *Longitude* relates to angular distance measured east or west of the prime meridian. *Latitude* is angular distance measured north or south from the plane established for the equator (figure 1.3). Other coordinate systems such as the Universal

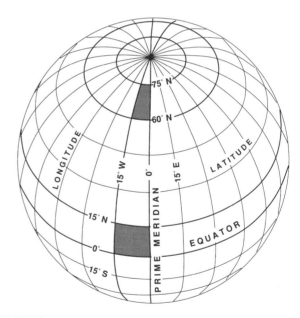

FIGURE 1.3

Longitude and latitude. The location of a place on the Earth is determined using longitude and latitude measurements. Each relates to angular distance measured on a spherical surface. Latitude is angular distance in a north-south direction; the equatorial plane (0°) is the starting point for such measurements. Longitude is angular distance in a west-east direction; the prime meridian (0°) is the base line used for longitude measurements. A whole series of longitude and latitude measurements provides the basis for creating the graticule or grid of the map.

Transverse Mercator (UTM), State Plane Coordinate System (SPCS), or national coordinate systems may be of assistance in defining the position of the study area and aiding in the location of appropriate base maps. By precisely defining the location of the map area, both politically and mathematically, a more systematic search for map sources may be conducted.

In a time when map libraries are installing data-base programs to manage their collections, a major locator is geographic coordinates. If you know the longitude and latitude of a place, a search can be made for all maps filed by that coordinate area. However, a search conducted in this manner may produce a listing of available products so long that it may be detrimental to the search process. When using a data base, you should include a combination of information such as coordinates and a geographic region to restrict the listing to only the most useful materials.

Knowledge of coordinates also promotes better use of index maps when collections are only cataloged by this form of reference.

Scale

The second step in identifying source materials is to determine the map scale. Map scale is a ratio, such as 1:24,000, which indicates how representative the map is of the area it is depicting. Ideally, the ratio would be 1:1, where one unit of map space would represent one unit of ground space (for example, one foot on the map would represent one foot on the ground). Because this is an impractical relationship, much larger ratios must be used. Typically, map scale is cited on the map as a representative fraction, such as 1:24,000. Here, one unit on the map (for example, inches or centimeters) corresponds to 24,000 units on the ground. Three different formats for expressing scale are shown in figure 1.4.

Frequently, scales are referred to as "small, medium, or large." These terms are confusing, because as the denominator of the representative fraction becomes *larger*, the scale actually becomes *smaller*. An easy way to visualize whether one map has a smaller scale than another is to imagine a ruler drawn to scale on

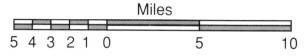

WRITTEN SCALE
One Inch Equals Five Miles

REPRESENTATIVE FRACTION
1: 316,800

BAR SCALE
Miles

5 4 3 2 1 0   5   10

FIGURE 1.4
Types of map scales. Three ways to express map scale are shown: the written scale, the representative fraction or ratio, and the graphic or bar scale. Scale defines how much work the map is doing to show the extent and location of cultural and natural features in the area being represented. For example, a large-scale map (such as 1:24,000) is able to show the position of most features more accurately than is the case for a small-scale product (such as 1:250,000).

each map. If each ruler represented the same distance on the ground, the smaller ruler would appear on the smaller-scale map. For example, if map "A" depicted a "mile-long ruler" as four inches long and map "B" depicted the same "mile-long ruler" as six inches long, then map "A" would be smaller in scale than map "B."

The scale of a map is often directly related to the quality and quantity of information depicted on that map. As noted earlier, most map compilations utilize source materials drawn at a larger scale. Seldom is compilation based on smaller scales, because of the danger of amplifying errors inherent in the smaller-scale products. At the very least, map compilation should be done with materials that are the same scale as the new map.

Timeliness of Source Material

A third concern is timeliness of the source materials; this always must be considered. When evaluating the time factor, especially for topographic maps, it is important to remember that *neither the date of publication nor the time of compilation is an accurate index to the currency of data displayed by the map.* In the case of topographic and related large-scale products, the best indicator is the time when aerial or satellite imagery was recorded. If the map has been updated, then the date of photo revision or the date of addition of survey data derived from the field provides a good guide to recency. For a thematic map, you should consider the date displayed in the title or in a map note as a reasonable means of evaluating the currency of statistical data.

Subject Material

A fourth step in resource evaluation is based upon the subject matter featured on the map. At some point in the evaluation process, you must relate the requirements of the new product to the content of previously published materials. For example, if the new map is to contain topographic information as part of its communication objective, then you must locate source maps that have the topographic data necessary to construct a terrain base for the new product. On the other hand, if the new map is to feature colonial economic activities such as mills, furnaces, ordinaries, or transportation, then the maps sought should display these features.

NATIONAL SOURCES FOR MAPS AND IMAGERY

For most mapping projects, you must find places where a search can be conducted for reference maps. Map collections may be maintained by geography, geology, and history departments at state colleges or universities. Some collections may be maintained by city, county, or state agencies, particularly tax and highway departments. Federal offices often are excellent sources for maps. Many libraries, public and college, are sources. All of these collections should be investigated in the research process.

You often will not find the necessary materials, and consequently will have to consult collections maintained at key federal agencies at the national level. Brief descriptions of three major sources follow: the Library of Congress,

the National Archives, and the National Cartographic Information Center. (Note: a list of major sources appears later in this section.)

Library of Congress

A particularly valuable source of geographic and cartographic information is the Geography and Map Division of the U.S. Library of Congress in Washington, DC (figure 1.5). This division is responsible for maintaining over 4,000,000 maps and 51,000 atlases. The collection is especially important as a source of foreign topographic base maps. Since 1968, new additions to the collection have been cataloged as part of the library's computerized data base. This cataloging data base is available to other libraries on magnetic tape (*MARC Maps*) or as 105 mm diazo microfiche (*National Union Catalog, Cartographic Materials*). Some universities and government agencies have these materials, and you should determine if access to them is available locally. More information about the collection is provided by Ronald Grim in this section's cameo essay.

National Archives

Another valuable source is the Cartographic and Architectural Branch of the National Archives and Records Service in Washington, DC. This branch is responsible for maintaining aerial photographs and manuscript maps produced by military and civilian agencies of the federal government. Many of the maps in this collection are original documents which provide significant insight into the history of our country and its portrayal on maps.

An interesting example is the set of maps depicting the construction of the Mullan Road in the Pacific Northwest (figure 1.6). This military wagon road, constructed during the years 1859–1862, connected Ft. Walla Walla in Washington Territory with Ft. Benton in Nebraska Territory. Examination of these maps reveals many interesting details about the construction of the road and illustrates the transition of terrain representation from hachures to contours in the nineteenth century. (For further information, see McDermott and Grim 1976, p. 221–226.)

FIGURE 1.5
Map reading room (**top**) at Library of Congress/Geography and Map Division. Researchers frequently visit the Map Reading Room to examine maps and other geographic materials stored in this part of the library. In the stack area (**bottom**), a large amount of floor space is devoted to the library's collection of over 4,000,000 maps. In addition, special file space is used to store atlases, globes, and terrain models.

FIGURE 1.6
Mullan Road drawing and topographic map. The Mullan Road was a military road constructed in the late 1850s to connect Ft. Walla Walla in Washington Territory to Ft. Benton in Nebraska Territory. It was about 600 miles long. The drawing (**top**) illustrates the road at a point west of Missoula, Montana (drawing by Gustavus Sohon, an early western artist). The map (**bottom**) is a portion of one of the Mullan Road maps. It is one of the first series of cartographic drawings to use the contour line as a major terrain symbol. (Both National Archives)

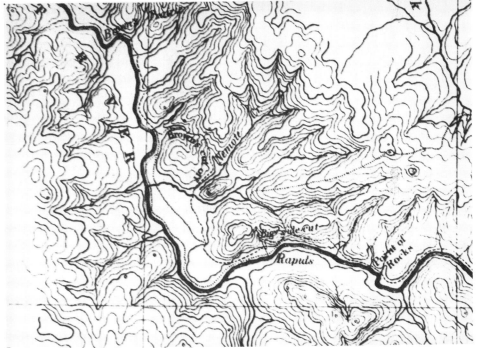

National Cartographic Information Center (NCIC)

A third major source of information is the National Cartographic Information Center (NCIC), part of the USGS. Currently, NCIC is located at the USGS national headquarters in Reston, Virginia, and regional and state-affiliate offices exist. You can contact NCIC for information about many different types of cartographic and photographic products produced by USGS and other federal agencies. The office also fills orders for special-purpose maps (such as land use and land cover), geographic-name lists, and standard topographic maps. Some of this material is indexed in a computer data base, part of which includes the Library of Congress collection. At this single location, you can obtain both valuable information and actual products.

Satellite Imagery

From time to time you will require satellite imagery for mapping projects. During 1986, the United States government sold its well-known and valuable civilian satellite system, LANDSAT-EROS, to the Earth Satellite Corporation (EOSAT) and the RCA Corporation. EOSAT, located in Washington, DC, is now responsible for providing remote-sensing imagery to civilian users. In addition, the French government is sponsoring a satellite system called SPOT. Their satellite, launched in February 1986, and is now providing high-quality imagery at a ground resolution of 10 meters (B/W) to 30 meters (color) (figure 1.7). You may contact these sources for hard copy and computer tapes covering their research areas. Addresses of these sources appear later in this section.

Index Sheets

Once you have identified the originating agency for a particular product, you can often request an index map that displays the geographic area covered by one or more map series. The outline of each map available in the series is shown on the index map. Figure 1.8 illustrates the new USGS design for a typical index map, and shows a portion of the coverage for Missouri and Illinois. Notice that each quadrangle map in the series has been referenced by name, correlated to an alphanumeric index on both large locator maps. (The reference codes used for identifying map products are explained in figure 1.9.)

Once you have located and identified the maps you need, you can create a list of map names for a given series, such as those with a scale of 1:24,000. You can then use these names either for ordering maps from the USGS, or for locating the maps in a collection. Index-map systems are also used for locating remote-sensing imagery produced by LANDSAT and SPOT.

Examination of index maps, especially for the larger-scale map series, usually indicates voids in coverage. Sometimes maps are unavailable, or they may be of inadequate quality for the new project. Voids of coverage are especially noticeable for foreign areas. Occasionally these deficiencies can be corrected by obtaining satellite imagery or by using smaller-scale materials and amplifying their information. On-site investigation can be made to acquire needed detail if time and money are available.

OBTAINING AERIAL PHOTOGRAPHS

Today's cartographer has access to millions of aerial photographs for gathering information. These images have been acquired through the combined efforts of both private and governmental agencies. The photography is usually available in different scales, sizes, types, and quality. Consequently, one of the major problems you will confront is determining what is available and from which agency the material can be acquired.

In general, the most economical source of aerial photographs is the federal government. For some very specialized applications, it is possible that a private group may have to be contacted for a suitable product. A good example is forensic use: civil and criminal trials frequently require timely, large-scale images to document the scenes of crimes or accidents.

The task or problem being undertaken will usually determine the sources or agencies you should contact. Once the source is selected, you must choose the final images to be acquired,

FIGURE 1.7
Portion of a SPOT satellite image. The harbor region of Baltimore, Maryland, is shown.
This black-and-white image has a ground resolution of 10 meters, meaning that under
ideal conditions, it is possible to see and perhaps identify any feature larger than 33
feet across. (Provided courtesy of SPOT Image Corporation, Reston, Virginia. Used by
permission. © 1987 by Centre National d'Etudes Spatiales)

38090

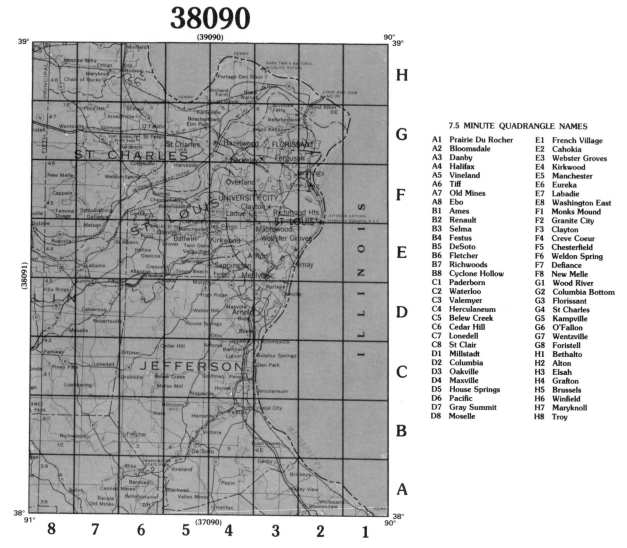

FIGURE 1.8

USGS index maps. Shown is the new-style format used in map catalogs being developed for the states. The name for a quadrangle is derived using an alphanumeric system. For example, in the table of quadrangle names, A1 is "Prairie Du Rocher." The two grid maps cover the same area; the left one shows detail for Missouri and the right one shows Illinois. (USGS)

considering such primary factors as the scale of the photography, the flight dates, and the amount of cloud cover. Other significant variables influencing image selection are film type, image cost, and availability of stereoscopic coverage.

Aerial Photography Summary Record System (APSRS)

Many government and private agencies produce aerial photographs and related remote-sensing images. You need to know what an agency has produced, and the character of its coverage. Obtaining this data is eased by APSRS, or the Aerial Photography Summary Record System. It describes the available aerial photography in the United States and related regions. The system is maintained by the U.S. Geological Survey and can be accessed through National Cartographic Information Center offices. APSRS is continually updated to provide information about the most recent available photography. APSRS information is provided

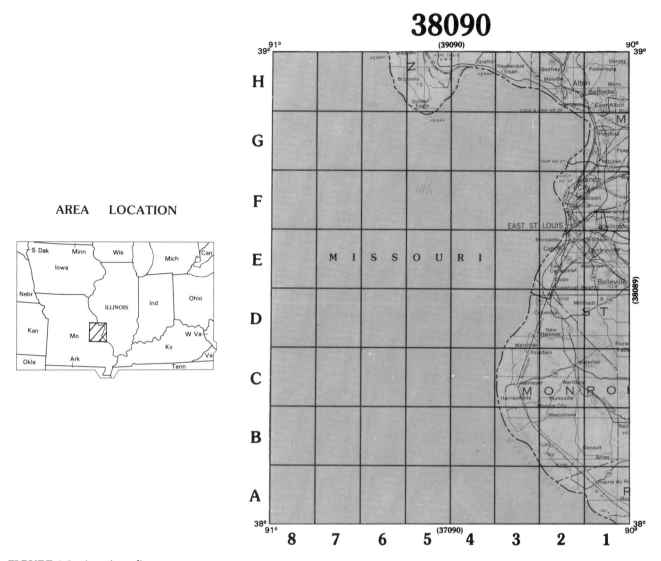

38090

AREA   LOCATION

FIGURE 1.8   (continued)

in three different formats—microfiche, summary maps, and custom queries.

Figure 1.10 shows the layout of a typical record, which is available either on microfiche or as a computer printout. The data is laid out in a columnar form with the subject of each column identified at the top of the sheet. Figure 1.11 explains the codes used on the APSRS sheets.

You may obtain information about availability of imagery from NCIC. One method is to complete a photographic coverage request form like that illustrated in figure 1.12, and mail it to the NCIC address shown under "Sources," later in this section. From the information provided, an NCIC employee will respond with a statement of photographs available and their cost. A second method is to determine the longitude and latitude of the location you are studying, and ask NCIC to send a printout of the summary record for that location. From this information you can select the images to acquire.

Private Sources of Aerial Photography

Private sources of imagery may not maintain computerized data bases for their products. Thus it may be necessary to visit the site where their records are maintained, and to describe the location, the time of year during which the image was made, the scale, and film type. From

EXPLANATION OF USGS MAP REFERENCE CODE

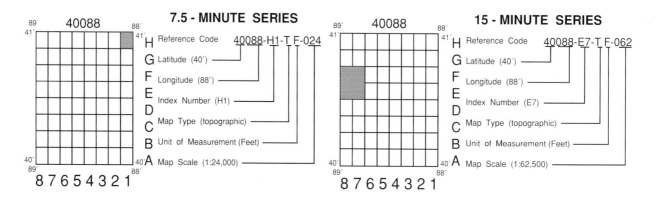

7.5 - MINUTE SERIES

H Reference Code — 40088-H1-T F-024
G Latitude (40°) ——
F Longitude (88°) ——
E Index Number (H1)
D
C Map Type (topographic)
B Unit of Measurement (Feet)
A Map Scale (1:24,000)

15 - MINUTE SERIES

H Reference Code — 40088-E7-T F-062
G Latitude (40°) ——
F Longitude (88°) ——
E Index Number (E7)
D
C Map Type (topographic)
B Unit of Measurement (Feet)
A Map Scale (1:62,500)

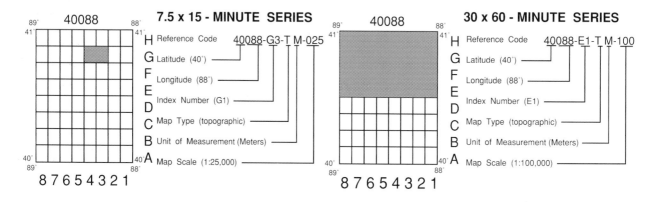

7.5 x 15 - MINUTE SERIES

H Reference Code — 40088-G3-T M-025
G Latitude (40°) ——
F Longitude (88°) ——
E Index Number (G1)
D
C Map Type (topographic)
B Unit of Measurement (Meters)
A Map Scale (1:25,000)

30 x 60 - MINUTE SERIES

H Reference Code — 40088-E1-T M-100
G Latitude (40°) ——
F Longitude (88°) ——
E Index Number (E1)
D
C Map Type (topographic)
B Unit of Measurement (Meters)
A Map Scale (1:100,000)

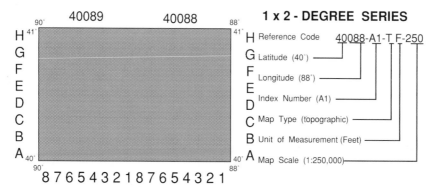

1 x 2 - DEGREE SERIES

H Reference Code — 40088-A1-T F-250
G Latitude (40°) ——
F Longitude (88°) ——
E Index Number (A1)
D
C Map Type (topographic)
B Unit of Measurement (Feet)
A Map Scale (1:250,000)

Portion of a Sample Order Form

FILE NO.	QTY	MAP NAME	DATE	REFERENCE CODE
00909		Abingdon	82	40090-G4-TF-024-00
00002		Adair	47	40090-D4-TF-024-00
00003		Addieville	74	38089-D4-TF-024-00
00004		Akin	63	37088-H6-TF-024-00
00005		Albion North	71	38088-D1-TF-024-00
			71	38088

FIGURE 1.9
Explanation of USGS Map Reference Code. Each map in a series has its own reference code, made up of a series of numbers and letters. The first number defines the latitude and longitude of the lower right corner of the map (40088, or lat. 40° N, long. 88° W). The next letter and number define the location of the map on the index map (H1). A letter is then used to define the map type (T for topographic). The unit of measurement used in the construction of the map is stated (F for feet, M for Meters). The final number denotes the map scale (250 for 1:250,000; 100 for 1:100,000). (USGS)

this information, the firm's staff will select photomosaics and/or roll film for you to examine. From this perusal, you can select images. You must specify the photographic medium (paper print, film positive, or transparency, etc.) and indicate the type of surface (glossy, matte) and weight (single, double) when prints are requested.

While there are too many private sources of aerial photography to list in this book, you can identify the principal companies that offer aerial photographic services by their advertisements in periodicals such as *Photogrammetric Engineering and Remote Sensing* and *Surveying and Mapping*. Also, consult the telephone directory serving your study area for companies offering aerial-photography services.

Sometimes it is impossible to find the required photography. Under these circumstances you may be forced to contract for a special flight to obtain the proper photography. Unfortunately, this is expensive, and the project may have to be deferred or even abandoned due to the cost.

SATELLITE IMAGERY

Annotation of LANDSAT Imagery Information describing the imaging parameters and characteristics appears on the margins of the imagery. A series of numbers identifies the longitude or latitude of each tick mark around the edges of the imagery. In the lower margin is the "annotation," a string of alphanumeric characters. A typical annotation appears in figure 1.13, with an explanation.

Image Date The characters **06APR77** indicate the date that the image was recorded: April 6, 1977.

Image Center The characters following the letter **C** define the center of the image in longitude and latitude. In the example, **N43-09** represents 43°9' north latitude and **W088-26** indicates 88°26' west longitude. If the letters were **S** and **E**, they would indicate south latitude and east longitude, respectively.

Path Identifier Letter **D** notes that the satellite was descending—moving from north to south. (An **A** would mean the satellite was ascending—south to north.) The numbers **025-030** following the letter denote the path (25) and row (30) position for the image. (On some images this field identifier may be allocated to specifying the originating agency, such as **USGS EDC**—USGS/Eros Data Center.)

Nadir The next sequence of letters and numbers beginning with **N** defines the nadir position of the satellite. In this example, **N43-02/ W088-14** means north 43°2' and west 88°14'.

Sensors and Spectral Bands The letters and numbers in this field define the sensor and bands used to make the image, and how the data was transmitted. **M** denotes that the image was multispectral (R would indicate that the return-beam vidicon system was used). The number indicates the band used to acquire the data (**7** refers to band 7). **D** means that the information was transmitted directly; **R** would signify that the data was stored and then played back from the satellite's RBV recorder.

Sun Angle The sun's angular position to the image is expressed in the sun's elevation above the horizon as measured from the center of the image, and the azimuth measured from true north. The annotation **SUN EL42 A131** positions the sun 42° above the horizon at the midpoint of the image, with an azimuth of 131° as measured from true north.

Geometric Correction, Processing, and Transmission This sequence of letters and numbers describes the geometry and transmission of the imagery.

*** C10 ***
NCIC

AERIAL PHOTOGRAPHY SUMMARY RECORD SYSTEM

INDEX CROSS REFERENCE

*** C10 ***
DATE: 02/05/83
PAGE 407

AGENCY CODE	RPT TYP	Q/W	SE CORNER LAT DEG MIN	SE CORNER LONG DEG MIN	FIPS CODE ST CTY	DATE OF COVERAGE YR MO DAY	STA	AGENCY PROJECT CODE	IMAGE SCALE	FOCAL LENGT	FILM TYPE	FILM FMT	SENS CLAS	CLOUD COVER	CAM SPEC	QUAD COVER	REMARKS OR SCENE ID	FRAMES FROM	FRAMES TO	CASSETTE NO.	FRAME
USAF	3	2	59 00	135 37		72 07 03	3	07140	65966	04	2	3	2	0		8	2 0223911				
USAF	3	2	59 00	135 37		72 07 03	3	07140	65934	04	2	3	2	0		8	2 0223861				
USAF	3	2	59 00	135 37		72 07 03	3	07140	66006	04	2	3	2	0		5	2 0223905				
USAF	3	2	59 00	135 37		72 07 03	3	07140	65949	04	2	3	2	0		5	2 0223869				
USAF	3	2	59 00	135 37		72 07 03	3	07140	65285	04	2	3	2	0		3	2 0223919				
USAF	3	2	59 00	135 37		72 07 03	3	07140	65168	04	2	3	2	0			2 0223886				
USAF	3	2	59 00	135 37		72 07 03	3	07140	65813	04	2	3	2	0			2 0223851				
USAF	3	2	59 00	135 37		71 06 03	3	07140	78990	04	2	3	2	0		6	2 0010050				
USAF	3	2	59 00	135 37		71 06 03	3	07140	79604	04	2	3	2	0		4	2 0010134				
USAF	3	2	59 00	135 37		71 06 03	3	07140	78677	04	2	3	2	3		2	2 0010078				
USAF	3	2	59 00	135 37		71 06 03	3	07140	78927	04	2	3	2	0			2 0010085				
USAF	3	2	59 00	135 37		71 06 03	3	07140	78039	04	2	3	2	0			2 0010098				
USN	3	2	59 00	135 37		50 09 21	3	SEW00	44000	04	4	3	1	0		2	2 0840048				
USN	3	2	59 00	135 37		48 08 14	3	SEA00	40000	04	4	3	1	0		7	2 1260019				
USN	3	2	59 00	135 37		48 07 17	3	SEA00	40000	04	4	3	1	0		2	2 0840048				
USN	3	2	59 00	135 37		48 07 05	3	SEA00	40000	04	4	3	1	0		4	2 0590164				
USN	3	2	59 00	135 37		48 06 05	3	SEA00	40000	04	4	3	1	0		8	2 0700100				
NASAJS	3	2	59 07	135 37		79 08 11	3	4060	122779	04	4	3	2	0		5	640600100	0190	0209	4061	0503
NASAJS	3	2	59 07	135 37		79 08 11	3	4060	61113	06	2	3	2	0		2	640600110	0212	0240	4062	0446
NASAJS	3	2	59 07	135 37		79 08 11	3	4060	121349	04	4	3	2	0			640600100	0114	0121	4061	0427
NASAJS	3	2	59 07	135 37		79 08 11	3	4060	60318	06	2	3	2	0			640600110	0001	0036	4062	0235
NASAJS	3	2	59 07	135 37		79 08 11	3	4060	121184	04	4	3	2	0			640600100	0001	0020	4061	0314
BLM	3	2	59 07	135 37		79	3	79-2	120000	04	4	3	1	0	Y	4	R-10	0062-0119			
BLM	3	2	59 07	135 37		79	3	79-4	60000	04	2	3	1	0	Y	3	R-11	0334-0421			
BLM	3	2	59 07	135 37		79	3	79-2	60000	04	2	3	1	0	Y	3	R-11	0122-0237			
BLM	3	2	59 07	135 37		79	3	79-1	60000	04	2	3	1	0	Y		R-11	0004-0119			
BLM	3	2	59 07	135 37		79	3	79-1	120000	04	4	3	1	0	Y		R-10	0003-0060			
NOS	3	2	59 07	135 37		74 08 01	3	74E-3	30000	04	4	3	1	0	Y	3	F4	6267-6275			
NOS	3	2	59 07	135 37		74 08 01	3	74E-2	30000	04	4	3	1	0	Y	3	F4	6265-6266			
NORPAS	3	2	59 07	135 37		74 07 08	3	GLABAY	76000	04	4	3	1	0	Y		GLACIER BAY				
NORPAS	3	2	59 07	135 37		74 07	3	74-2	76000	04	4	3	1	0	Y	9	GLACIER BAY-AK				
USAF	3	2	59 07	135 37		72 07 03	3	07140	66006	04	2	3	2	0		9	2 0223905				
USAF	3	2	59 07	135 37		72 07 03	3	07140	65949	04	2	3	2	0		5	2 0223869				
USAF	3	2	59 07	135 37		72 07 03	3	07140	65966	04	2	3	2	0		4	2 0223911				
USAF	3	2	59 07	135 37		72 07 03	3	07140	65342	04	2	3	2	0		3	2 0223926				
USAF	3	2	59 07	135 37		72 07 03	3	07140	65168	04	2	3	2	0			2 0223886				
USAF	3	2	59 07	135 37		72 07 03	3	07140	65636	04	2	3	2	0			2 0223900				
USAF	3	2	59 07	135 37		72 07 03	3	07140	65934	04	2	3	2	0			2 0223861				
NOS	3	2	59 07	135 37		71 06 05	3	71E-2	30000	04	4	3	1	0	Y	4	F4	4865-4874			
NOS	3	2	59 07	135 37		71 06 05	3	71E-1	30000	04	4	3	1	0	Y	2	F4	4856-4863			
USAF	3	2	59 07	135 37		71 06 03	3	07140	78990	04	2	3	2	0		6	2 0010050				
USAF	3	2	59 07	135 37		71 06 03	3	07140	79604	04	2	3	2	0		5	2 0010134				
USAF	3	2	59 07	135 37		71 06 03	3	07140	78939	04	2	3	2	0		5	2 0010094				
USAF	3	2	59 07	135 37		71 06 03	3	07140	78039	04	2	3	2	0			2 0010098				
USAF	3	2	59 07	135 37		71 06 03	3	07140	78927	04	2	3	2	0			2 0010085				
NOS	3	2	59 07	135 37		66 06 28	3	66L-2	30000	04	4	3	1	0	Y	3	F4	5329-5337			
USN	3	2	59 07	135 37		48 08 25	3	SEA00	40000	04	4	3	1	0		3	2 1400096				
USN	3	2	59 07	135 37		48 08 14	3	SEA00	40000	04	4	3	1	0		9	2 1260058				
USN	3	2	59 07	135 37		48 08 13	3	SEA00	40000	04	4	3	1	0		4	2 1150086				
USN	3	2	59 07	135 37		48 07 17	3	SEA00	40000	04	4	3	1	0		5	2 0840141				
USN	3	2	59 07	135 37		48 07 04	3	SEA00	40000	04	4	3	1	0		6	2 0660080				
USN	3	2	59 07	135 37		48 07 04	3	SEA00	40000	04	4	3	1	0		5	2 0660106				
USN	3	2	59 07	135 37		48 07 04	3	SEA00	40000	04	4	3	1	0		2	2 0670040				

FIGURE 1.10
A record from the Aerial Photography Summary Record System (APSRS). Whether microfiche or computer printout, data regarding the availability of aerial photography is provided in a format like the one shown. (USGS)

The first three characters identify the kind of geometric correction, scale, and projection used in creating the imagery (**S1S** in the example). The first character, geometric correction, can be **U** = uncorrected, **S** = system level, **G** = geometrically corrected based on geometric GCPs (ground control points), or **R** = geometrically corrected based on relative GCPs. The second character, scale of imagery, is indicated by **1** = 185 km × 185 km (100 nm × 100 nm), **2** = 92.5 km × 92.5 km (50 nm × 50 nm), or **3** = 185 km × 170 km (100 nm × 92 nm). (The abbreviation nm = nautical miles, which is approximately 6076 feet.) The third character, image projection, can be **L** = Lambert, **P** = polar stereographic, **H** = hotline oblique, **S** = space oblique Mercator, **U** = universal transverse Mercator, or **N** = natural perspective.

Following the hyphen is information relating to the resampling algorithm and ephemeris data used to compute the image. The first character indicates the resampling algorithm: **C** = cubic, **N** = nearest neighbor. The second character signifies the ephemeris data: **P** = predictive, **D** = definitive.

After the hyphen come the processing and transmission annotations. The first character in this sequence indicates the type of processing procedure: **N** = normal processing procedure, **A** = abnormal processing procedure. The second character indicates whether an Earth image or an RBV (return beam vidicon) image has been processed: a blank = earth image; **0, 1,** or **2** = RBV radiometric calibration images indicating lowest-to-highest exposure levels, respectively. In the example, a blank appears because the imagery is MSS (multispectral system), not RBV. The third character signifies the sensor gain options: **H** = high gain, **L** = low gain. The final character in the sequence

EXPLANATION OF THE CODES USED IN THE
AERIAL PHOTOGRAPHY SUMMARY RECORD SYSTEM

AGENCY CODE
Identifies the agency to contact to acquire specific information about the photography and/or to obtain copies or use of the photography.

RPT TYPE (Report Type)
1 = county format
2 = 7.5' quad format
3 = four corner format

Q/W (Quadrant of the World)
1 = north lat. and east long.
2 = north lat. and west long.
3 = south lat. and east long.
4 = south lat. and west long.

SE CORNER (Lat/Long -- Deg/Min)
Degree and minute of latitude and longitude of southeast corner of 7.5' quadrangle.

FIPS CODE (State/County)
State and county numbers assigned using Federal Information Processing Standards Publication Codes.

DATE OF COVERAGE (YR/MO/DAY)
Year, month and day the photography was completed.

STA (Status)
3 = photography completed

AGENCY PRODUCT CODE
Project identification assigned by flying agency.

IMAGE SCALE
Scale of imagery expressed as a whole number.

FOCAL LENGT (Focal Length)

01 = 1.75" or 44mm
02 = 3.00" or 76mm
03 = 3.46" or 88mm
04 = 6.00" or 152mm
05 = 8.25" or 210mm
06 = 12.00" or 305mm
07 = 24.00" or 610mm
08 = 3.96" or 100mm
09 = 9.430" or 240mm
10 = 6.738" or 171mm
11 = 3.35" or 85mm
12 = 8.11" or 206mm
13 = 1.38" or 35mm
14 = 1.97" or 50mm
15 = 3.94" or 100mm
16 = 4.13" or 105mm
17 = 5.31" or 135mm
18 = 5.91" or 150mm
19 = 9.84" or 250mm
20 = Other

FILM TYPE (Emulsion)
1 = W&W IR
2 = Color IR
3 = Color
4 = B&W
5 = Thermal
6 = Other

FILM FMT (Film Format)
1 = 2.76" or 70mm
2 = 4.5" x 4.5" or 11cm x 11cm
3 = 9" x 9" or 23cm x 23cm
4 = 9" x 18" or 23cm x 46cm
5 = 7" x 7" or 18cm x 18cm
6 = 7" x 9" or 18cm x 23cm
7 = 6" x 8" or 15cm x 20cm
8 = 1.0" x 1.38" or 25mm x 35mm
9 = Other

SENS CLAS (Sensor Class)
1 = Vertical carto (Implies Stereo)
2 = Vertical reconnaissance
3 = SLAR
4 = Oblique
5 = Other

CLOUD COVER (Percentage of)
0 = 0%   4 = 40%   7 = 70%
1 = 10%   5 = 50%   8 = 80%
2 = 20%   6 = 60%   9 = 90%
3 = 30%

CAM SPEC (Camera Specifications)
Indicates if camera meets calibration specifications.
Y = Yes   N = No   Blank = Unknown

QUAD COVER (Quad Coverage)
1 = 10%   5 = 50%   8 = 80%
2 = 20%   6 = 60%   9 = 90%
3 = 30%   7 = 70%   Space = 100%
4 = 40%

REMARKS
This is a dual-purpose field. The heading REMARKS usually refers to free-form data entered by agencies other than NASA. For example, the USFS frequently enters the name of a National Forest. The USGS usually enters QUAD-CENTERED on planned photographs and a microfilm identification on completed photographs.

| SCENE ID | FRAMES FROM TO | CASSETTE NO. FRAME |
The subheadings refer to NASA data only, and identifies the actual photographic frame containing the quadrangle coverage and microfilm cassette locator information.

FIGURE 1.11
Explanation of codes used in APSRS records. (USGS)

Name: _____

Street address: _____

City and State: _____

Telephone: (res): _____

(work): _____

Company affiliation: _____

Specific Area:

State: _____ County: _____

Town: _____

Marked map enclosed: ☐   Marked map enclosed: ☐

Section, township, and range (if known): _____

Geographic coordinates: _____

Feature that you specifically want to see: _____

Film:

Black-and-white: ☐   Color: ☐

Color-infrared: ☐   No preference: ☐

Date of photography:

Year: _____

Time of year, if you have a specific requirement:

Spring: ☐   Fall: ☐

Summer: ☐   Winter: ☐

Oldest available: ☐   Most recent available: ☐

No preference: ☐

Level of Desired Ground Detail:

Low altitude, large scale---Maximum detail: ☐

High altitude, small scale---Minimum detail: ☐

Size: (not available for all aerial photographs)

Standard contact: 9" x 9": ☐   2X (18" x 18"): ☐

3X (27" x 27"): ☐   4X (36" x 36"): ☐

Stereoscopic:

Do you want a series of overlapping photos for viewing through a stereoscope?

Yes: ☐   No: ☐

Purpose:

We can advise you on the best selection available if you describe what you intend to use the photographs for:

_____

_____

_____

FIGURE 1.12
Photographic coverage request form. By answering questions on the form, the cartographer defines the parameters used to select the needed photography. (USGS)

06APR77 C N43-09/W088-26 D025-030 N N43-02/W088-14 M 7 D
SUN EL42 A131 S1S-CD-N L1 NASA LANDSAT E-2 005-15411

FIGURE 1.13
LANDSAT annotation. (USGS)

identifies the transmission mode: **1** = linear mode, **2** = compressed mode.

Agency and Project In the example, **NASA LANDSAT** defines the originating agency and the project.

Image Identification Number The final sequence of characters defines the image identifier number for a scene. **E** indicates that the data is encoded. **2** means that the image was acquired from LANDSAT 2. **005** indicates the day of the observation relative to the launch.

The five-digit time annotation means: **15** = 15 hours, **41** = 41 minutes, **1** = 10 seconds (seconds are given in tens).

Acquisition of Satellite Imagery

Satellite imagery from LANDSAT and SPOT is available in a wide variety of formats, such as computer-compatible tapes (CCT), color composites, film positives and negatives, and paper prints. The type of format ordered depends on the image-analyzing equipment used and the type of problem being studied.

Commercialization of image-acquisition programs has led to the high cost of imagery. In addition, substantial fees are added to the image costs if the imagery is not "off the shelf" and must be made to your specifications of time and location. Another fee is added if you require the creation of an image with minimal cloud cover!

Ordering LANDSAT and SPOT Imagery

The key to obtaining imagery is specification of a particular place or site. To order imagery, you must establish longitude and latitude coordinates for the study area. These coordinates can then be related to the orbital path of the satellite. Once this relationship is established, you can request the imagery.

LANDSAT images are indexed by the Worldwide Reference System (WRS). WRS is a global network of paths and rows by which all scenes can be geographically located. Index maps showing availability of imagery are published for different regions of the world. A section of an index map for LANDSAT is illustrated in figure 1.31; the index shows the orbit paths of the satellite and the image rows.

Figure 1.14 illustrates the relationship of the LANDSAT scene to path and row numbers. The intersection of the center of a row with the orbital path marks the center of the image scene. You can translate the geographical coordinates of your study area to the nearest path and row. For LANDSATs 4 and 5, WRS yields 233 paths and 248 rows.

Once you have determined the path and row numbers, you can order an image, or locate it for previewing, using a catalog system. This catalog is organized by zones and the WRS rows and paths within each zone. Each zone is placed upon a microfiche which contains 60 microframes describing the image data for the row and path within a prescribed time period. Once this data has been perused, you can preview

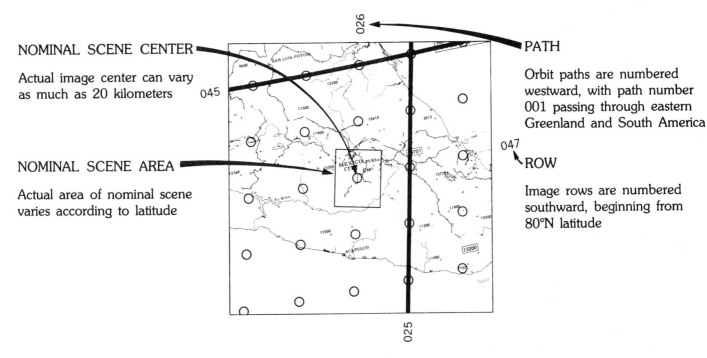

NOMINAL SCENE CENTER

Actual image center can vary as much as 20 kilometers

NOMINAL SCENE AREA

Actual area of nominal scene varies according to latitude

PATH

Orbit paths are numbered westward, with path number 001 passing through eastern Greenland and South America

ROW

Image rows are numbered southward, beginning from 80°N latitude

FIGURE 1.14
Explanation of LANDSAT image reference system. To determine the availability of satellite imagery, you must relate geographic location to particular row and path numbers that indicate the point at which an image was recorded. *Paths* mark the orbit lines followed by the satellite. *Rows* indicate the latitudinal location of the satellite at the time of a scan. (USGS)

SPOT Grid Reference System (GRS) Map of the United States

Earth resources data from the SPOT satellite are available from SPOT Image Corporation for areas around the world. These remotely sensed data, with 10 and 20 meter resolution, are produced in both digital and photographic forms at the SPOT Image processing facility in Reston, Virginia.

The SPOT satellite system was developed by France in cooperation with Belgium and Sweden, and is operated by the Centre National d'Etudes Spatiales (CNES). Imagery generated by SPOT is distributed by SPOT IMAGE, S.A. and its licensed distributors. SPOT Image Corporation is the exclusive distributor of SPOT imagery in the United States.

This map displays the SPOT Grid Reference System of the United States. It can be used to identify the appropriate column and row numbers for obtaining archived SPOT data or for requesting special acquisitions.

Orbital Parameters

The SPOT satellite is maintained on a near-polar, circular sun-synchronous orbit at an altitude of 832 km (at 45 degrees north); the mean orbital period is 101.4 minutes. The phased orbit enables the satellite to overfly each of the 369 ground tracks every 26 days, the duration of an orbital cycle. The satellite tracks are 108.6 km apart at the equator, and draw closer at higher latitudes. SPOT passes over the United States from north to south at about 11:00 a.m. (mean local solar time).

Satellite Imaging Parameters

The mirrors on each of the high resolution imaging instruments (HRVs) can be adjusted to view areas up to 475 km to either side of the satellite track. Generation of a SPOT scene (60 km long) takes about 9 seconds; two scenes can be generated at any one time. When the SPOT satellite is operated in the *twin vertical configuration,* the two identical HRVs are centered on the GRS nodal points to either side of the satellite track, yielding two 60 km-wide scenes with 3 km of sidelap; in effect, a 117 km-wide swath centered over the track. These nominal scene centers, or nodal points, are identified in the SPOT catalog by scene designators K (column) and J (row).

In the case of *oblique viewing,* the scene centers are located on a row J, but generally do not coincide with longitudinal positions of nodes on the GRS map. Scenes acquired by oblique viewing are referred to by the K, J values of the node nearest to the scene center.

Each satellite track is designated by two numbering systems: satellite track numbers (N) label the tracks from west to east from 1 to 369; revolution numbers (R) indicate the order in which each of the 369 tracks are actually overflown during an orbital cycle. Please note that the two column numbers (K), which are associated with a satellite track and referenced in the SPOT catalog are derived from the satellite track numbers (N).

Values for:
  K Column Number   **N** Satellite Track Number
  J Row Number   **R** Revolution Number
may be read directly from the map.

Zones

The earth is divided by the GRS into five latitudinal zones which have the following ranges:
- **North Polar** — north of 71.7° N
- **North** — **from 51.55° N to 71.7° N**
- **Intermediate** — **from 51.55° S to 51.55° N**
- **South** — from 51.55° S to 71.7° S
- **South Polar** — south of 71.7° S

The United States is located in the North and Intermediate zones.

The closely spaced tracks in northern latitudes allow for numerous opportunities to image a particular area. Because they would otherwise overlap, the GRS nodes in the **Northern** and **Southern** zones exist only for the *odd numbered* satellite tracks (N). Thus, in the northern zone, as in all areas of the world, to identify a scene acquired along an even numbered track, refer to the value of the nearest GRS node.

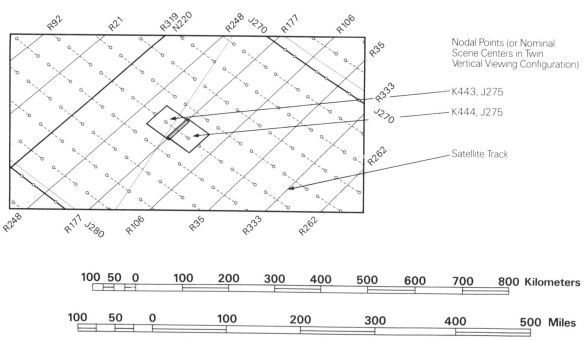

Nodal Points (or Nominal Scene Centers in Twin Vertical Viewing Configuration)

K443, J275
K444, J275
Satellite Track

| 100 50 0 | 100 | 200 | 300 | 400 | 500 | 600 | 700 | 800 Kilometers |

| 100 | 50 | 0 | 100 | 200 | 300 | 400 | 500 Miles |

Base map compiled by the U.S. Geological Survey (1975)
SPOT Reference Grid traced by the Institut Geographique National, France.

The GRS is described in greater detail in the SPOT User's Manual. GRS map sheets are available for many areas of the world. For copies of these maps or for more information on obtaining SPOT data, please contact: **SPOT Image Corporation**
1897 Preston White Drive
Reston, VA 22091-4326
(703) 620-2200
Telex 4993073 or your local distributor.

Lambert Conformal Conic Projection based on Standard Parallels 37° and 65° N
Scale 1:6,000,000

FIGURE 1.15
SPOT grid reference system. SPOT uses a slightly different alphanumeric system for identifying image locations. The system uses the letters K for path (orbit) positions and J for row (latitude) positions. An additional difference is in the plotting of points to indicate availability of stereo images. (Provided courtesy of SPOT Image Corporation, Reston, Virginia. Used by permission. © 1987 by CNES)

the most desirable scenes, using 16 mm black-and-white microfilm. If they are acceptable, you can order the desired images.

The SPOT indexing system is diagrammed and discussed in figure 1.15. This system is similar to the LANDSAT indexing system, but the terminology changes from *rows* and *paths* to *rows* and *columns*. In addition, the satellite *track number* and *revolution numbers* are included on the index sheet, because of the added capability of the SPOT system to acquire oblique imagery.

A set of WRS indexed maps portraying the areal coverage of row and path information is available from EOSAT (LANDSAT). To contact EOSAT or SPOT, please see "Sources" later in this section.

CAMEO

Dr. Ronald E. Grim

Bibliographer, Geography and Map Division, Library of Congress

Dr. Ronald E. Grim is Bibliographer in the Geography and Map Division of the Library of Congress. Prior to this appointment he was Assistant Chief for Reference in the Center for Cartographic and Architectural Archives, National Archives and Records Service, in Washington. He received his B.A. in history and geography from Muskingum College, New Concord, Ohio, and his M.A. and Ph.D. degrees in historical geography from the University of Maryland at College Park.

Introduction

Outside of conducting your own surveys, the major source of information in compiling a map is already-existing maps. In other words, map compilation, at least at the elementary stage, is accomplished by gathering information from secondary sources (i.e., maps that have already been compiled or published). Notwithstanding the problems of copyright and the repetition of cartographic errors made by others, which are separate problems in themselves, this essay identifies some major sources of cartographic information that will be useful for beginning cartographic students. Specifically, it focuses on the nature and availability of map collections; bibliographies or listings of maps, often referred to as cartobibliographies; and standard base maps useful in constructing topographic and thematic maps.

How to Find Important Map Collections

There are numerous map collections throughout the United States and Canada. Most universities and colleges with major geography and cartography programs have at least one map collection. In addition, map collections are found in state libraries, archives, or historical societies; major public city libraries; selected privately endowed libraries; and various national, state, and local government agencies concerned with mapping activities. The most recent guides to these collections are David K. Carrington and Richard W. Stephenson, eds., *Map Collections in the United States and Canada: A Directory*, 4th ed. (New York: Special Libraries Association, 1985) and Lorraine Dubriel, ed., *Directory of Canadian Map Collections*, 4th ed. (Ottawa: Association of Canadian Map Libraries, 1980). Another useful guide to map collections outside the United States and Canada is John A. Wolter, Ronald E. Grim, and David K. Carrington, eds., *World Directory of Map Collections*, 2d ed., International Federation of Library Associations and Institutions, Publication 31 (Munich: K.G. Saur, 1986).

The types of maps found in these collections vary according to the clientele and the acquisition policies of the respective institutions. The map holdings may range from current topographic map series of an entire state or country to historical or archival maps produced by a number of government agencies. The directories provide brief overviews of each institution's holdings, while more detailed listings may be available in the form of card catalogs, computerized data bases, or published cartobibliographies produced by the individual repositories.

A brief review of the holdings and finding aids of the two major map collections—the Geography and Map Division in the Library of Congress and the Cartographic and Architectural Branch in the U.S. National Archives—will provide an idea of the types of maps and services that can be found in other map collections which may be more accessible to you. These two collections, both located in the Washington, DC metropolitan area, utilize two different approaches to the acquisition and organization of cartographic materials: the first, a library approach; the second, an archival approach.

The Library of Congress Map Collection

The Geography and Map Division of the Library of Congress, with a collection comprising almost 4,000,000 maps and 51,000 atlases, is the largest map collection in the world. Although there is a strong emphasis on the United States during the eighteenth, nineteenth, and twentieth centuries, the geographic coverage is worldwide with examples of cartographic materials dating back to the late fifteenth and early sixteenth centuries. These cartographic materials have been acquired largely through copyright and government deposit, foreign exchange, gifts, and purchase. They are arranged according to a geographic classification scheme, facilitating the majority of reference requests, which focus on specific geographic localities.

There is no comprehensive catalog describing each individual item in the collection. Topographic and thematic map series from the United States and foreign countries, which comprise about 55% of the collection, are controlled with individual series descriptions and graphic indexes showing the coverage of each series.

Atlases acquired through the 1960s are described in the multivolume work, Philip Lee Phillips and Clara Egli LeGear, *A List of Geographical Atlases in the Library of Congress*, 8 vols. (Washington: Library of Congress, 1909–1974). Atlases pertaining to the United States are also described in two publications: Clara Egli Le-

Gear, *United States Atlases: A List of National, State, County, City, and Regional Atlases in the Library of Congress* (Washington: Library of Congress, 1950; reprint ed. New York: Arno Press, 1971) and Geography and Map Division, Reference and Bibliography Section, *Fire Insurance Maps in the Library of Congress: Plans of North American Cities and Towns Produced by the Sanborn Map Company* (Washington: Library of Congress, 1981). Figure 1.16 illustrates an example of a fire-insurance map. In addition, most atlases (except the fire-insurance atlases) have been cataloged and can be accessed through a card catalog and the computerized data base.

Single maps acquired since 1968 have also been cataloged and can be accessed through the computerized data base. Unfortunately, there is no similar access to single maps acquired before 1968. Single maps are arranged according to a geographical hierarchy (continent, country, state or province, county, and city). Maps pertaining to selected topics have been described in published bibliographies. Some of these publications include: Richard W. Stephenson, *Civil War Maps: An Annotated List of Maps and Atlases in Map Collections of the Library of Congress* (Washington: Library of Congress, 1961); Stephenson, *Land Ownership Maps: A Checklist of Nineteenth Century United States County Maps in the Library of Congress* (Washington: Library of Congress, 1975); and John R. Hebert and Patrick E. Dempsey, *Panoramic Maps of Cities in the United States and Canada: A Checklist of Maps in the Collections of the Library of Congress, Geography and Map Division*, 2d ed. (Washington: Library of Congress, 1984). Examples of Civil War and land ownership maps are seen in figures 1.17 and 1.18.

Cataloging information, which has been entered into the computerized data base, can be obtained by other libraries on subscription basis in the form of catalog cards, magnetic tapes (*MARC Maps*, which is distributed every four weeks), and 105 mm diazo microfiche (*National Union Catalog, Cartographic Materials*, which is issued quarterly) from the Library's Cataloging Distribution Service. Also available is a complete list of publications pertaining to the Geography and Map Division, and ordering instructions; these are available free from the Geography and Map Division.

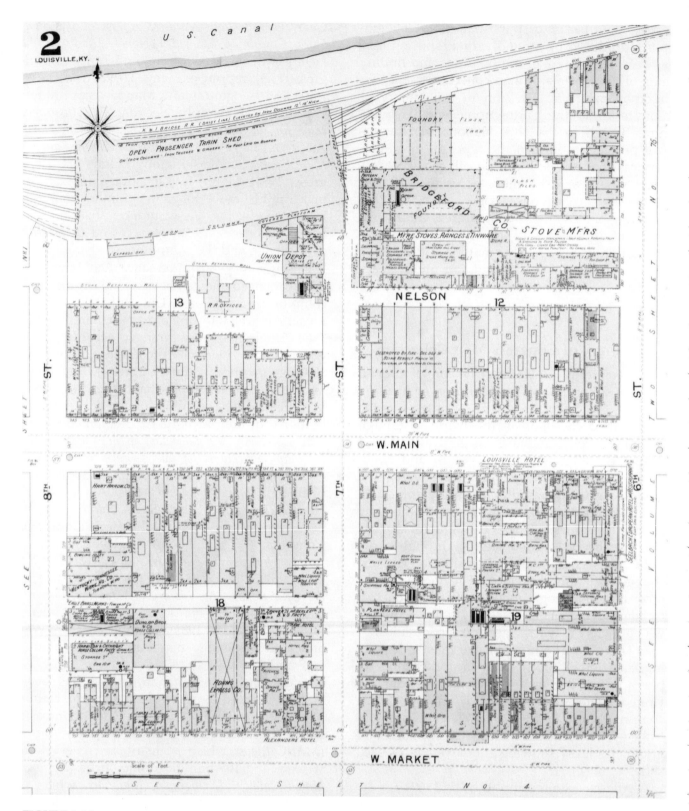

FIGURE 1.16
Section of nineteenth-century Sanborn Fire Insurance Map. These detailed maps have been used since the late 1800s to show the type of construction in various buildings in urbanized areas. This example shows part of Louisville, Kentucky, featuring the part of the city known for its iron-front buildings. Sanborn maps are frequently used as primary documents to ascertain land uses characteristic of different cities during the past 100 years. Adjacent page shows an enlarged portion of this figure. (Library of Congress)

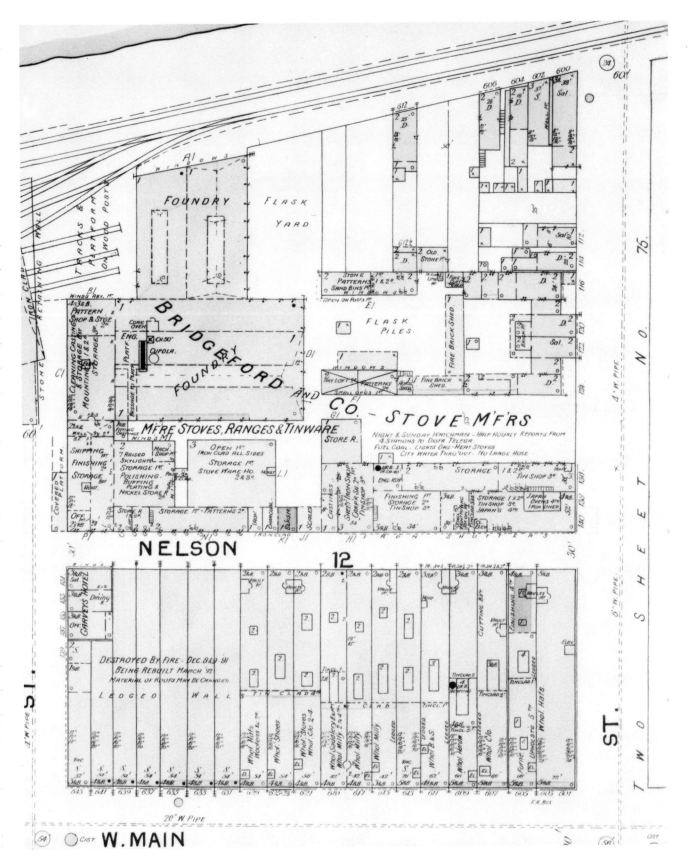

FIGURE 1.16 (continued)

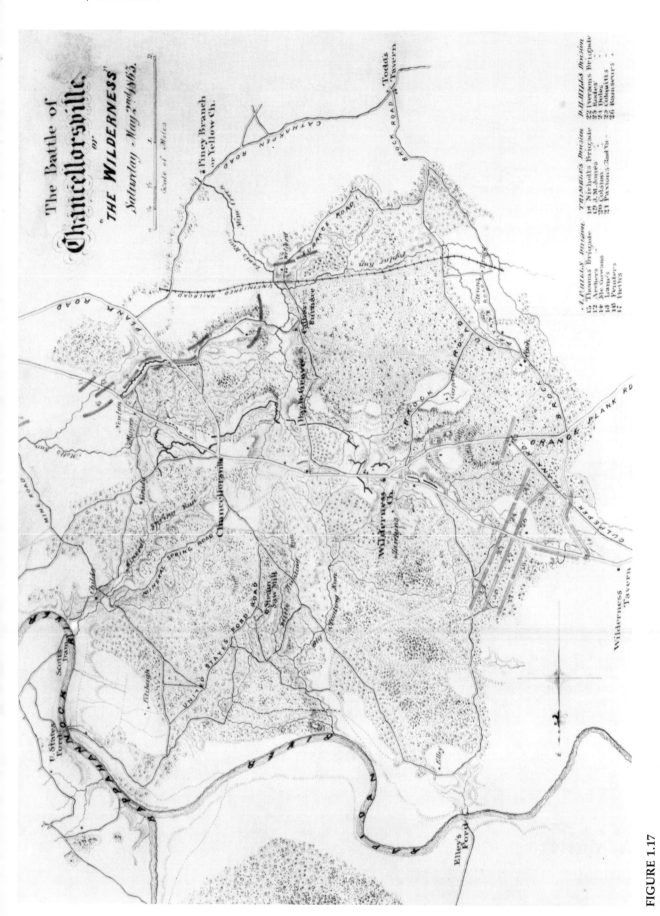

FIGURE 1.17
Civil War map. Among the maps most used by researchers at the Library of Congress and the National Archives are those featuring Civil War battle sites. This map illustrates troop positions at the famous battle of Chancellorsville or The Wilderness. Critical portions of the battle took place on May 2, 1863. (Library of Congress)

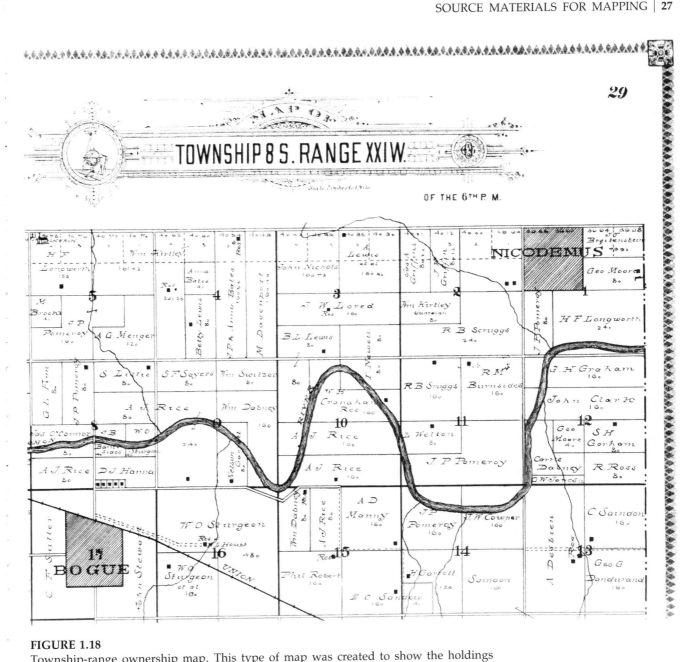

FIGURE 1.18
Township-range ownership map. This type of map was created to show the holdings of persons residing in the area covered by the product. Part of Township 8S, Range 21W is illustrated. This type of map is exploited in historical research to determine how the land was used, and how these uses have changed over time. Genealogists often use these drawings to map the location and migration of family members. (Library of Congress)

National Archives Collection

Although the cartographic collection in the Cartographic and Architectural Branch of the U.S. National Archives is only about half the size of that in the Library of Congress, it is the largest collection of archival maps and aerial photographs in the world. The collection numbers about two million maps, charts, and architectural drawings, and seven million aerial photographic images. It concentrates mainly on the United States during the nineteenth and twentieth centuries. Geographical coverage is worldwide, although it focuses primarily on foreign areas where the United States government has had an active interest during the last century.

The major differences between an archival map collection and a library map collection are the source of acquisitions and the method of organization. In the National Archives, maps and aerial photographs are acquired only from other agencies of the federal government. These materials were created by federal agencies but are no longer needed for current work. Many of the maps are manuscript (hand-drawn originals), although the Archives also maintains a set of published maps produced by each agency (this is the major area of overlap in the collections held by the Library of Congress and the National Archives).

The guiding principle in the arrangement of archival materials is according to "provenance" and "respect des fonds," meaning that the first level of arrangement is according to the agency that produced the records, and secondly the filing scheme created by the originating agency is maintained. Consequently, there are numerous self-contained map collections, each with its own filing scheme. This arrangement makes geographical research more difficult, but it facilitates an understanding of the map-creation process and allows the correlation of written documents that may explain the historical significance of the maps.

As in the Library of Congress, there is no comprehensive catalog of the individual maps in the collection. However, there is a comprehensive publication, *Guide to Cartographic Records in the National Archives* (Washington: National Archives and Records Service, 1971), which provides a brief collective description of each archival series of maps that was in the National Archives as of 1966.

Most series of maps are controlled by manuscript lists, graphic indexes, or catalog cards. In addition, published map bibliographies are available for selected "record groups" (the designation for the records from one agency or bureau) or selected topics. Some of these publications include *Civil War Maps in the National Archives* (Washington: National Archives and Records Service, 1964); Laura E. Kelsay, *Cartographic Records of the Bureau of Indian Affairs*, Special List No. 13, rev. ed. (Washington: National Archives and Records Service, 1977); Kelsay, *List of Cartographic Records of the General Land Office*, Special List No. 19 (Washington: National Archives and Records Service, 1964); Charles E. Taylor and Richard E. Spurr, *Aerial Photographs in the National Archives*, Special List No. 25, rev. ed. (Washington: National Archives and Records Service, 1973); and Janet L. Hargett, *List of Selected Maps of States and Territories*, Special List No. 29 (Washington: National Archives and Records Service, 1971). A complete list of publications is available free from the branch office.

These two map collections are not readily accessible to everyone, nor will everyone need the depth of coverage provided by their holdings. Many of the published maps held by these two institutions are available in local collections.

Cartobibliographies

To determine what maps have been published for a particular area or theme, and their availability in a particular collection, you will need to consult relevant cartobibliographies (map or atlas bibliographies). Although there is no comprehensive bibliography of cartobibliographies, there is one primary reference tool for cartobibliographic research: the seven-volume set *Bibliography of Cartography* (Boston: G.K. Hall, 1973–80), which is a product of the Library of Congress Geography and Map Division. It covers literature pertaining to cartography, the history of cartography, map librarianship, and related topics. Included in this dictionary listing are numerous cartobibliographies. In this reference are multiple entries for each citation, usually one under the name of the author or compiler, and several under pertinent subject headings.

Most bibliographical publications pertaining to maps are referenced under the name of a geographic area followed by "Maps—Bibliography." For example, a cartobibliography of Virginia maps is cited under the subject heading "Virginia—Maps—Bibliography." Another useful listing of recent cartobibliographical literature pertaining to the United States is found in Chapters 1–4, pages 3–51 of Ronald E. Grim, *Historical Geography of the United States: A Guide to Information Sources* (Detroit: Gale Research, 1982).

Although it is impossible to list here all the existing cartobibliographies pertaining to the United States, the following discussion of the more significant ones gives an idea of the types that have been published. One covering a fairly large geographic area is Robert W. Karrow, Jr., ed., *Checklist of Printed Maps of the Middle West to 1900*, 14 vols. in 12 (Boston: G.K. Hall, 1981–1983). This publication is actually a union list of printed maps published before 1900, indicating their location in one or more of the 127 repositories surveyed. Similar union lists have been published for Vermont, New Hampshire, and Utah: David A. Cobb, "Vermont Maps Prior to 1900: An Annotated Cartobibliography," *Vermont History* 39 (Summer and Fall 1971): 169–317; Cobb, *New Hampshire Maps: An Annotated Checklist* (Hanover: University Press of New England for New Hampshire Historical Society, 1981); and Riley M. Moffat, *Printed Maps of Utah to 1900: An Annotated Cartobibliography*, Western Association of Map Libraries Occasional Paper No. 8 (Santa Cruz, CA: Western Association of Map Libraries, 1981).

Some cartobibliographies pertain to a specific time period or subject. For example, James C. Wheat and Christian F. Burn, *Maps and Charts Published in America before 1800: A Bibliography* (New Haven: Yale University Press, 1969; rev. ed., London: Holland Press, 1978) attempts to list all the printed maps published before 1800. Maps of the Revolutionary War are listed in Kenneth Nebenzahl, *A Bibliography of Printed Battle Plans of the American Revolution, 1775–1795* (Chicago: University of Chicago Press, 1975).

Other guides may list only maps found in one institution. Besides those already mentioned for the Library of Congress and the National Archives, other examples include Martha L. Simonetti, *Descriptive List of the Map Collection in the Pennsylvania State Archives* (Harrisburg: Pennsylvania Historical and Museum Commission, 1976), and James Day, et al., *Maps of Texas, 1527–1900: The Map Collection of the Texas State Archives* (Austin: Pemberton Press, 1964).

Obviously, cartobibliographies will be of varying utility, depending on their scope and level of description. Nevertheless, a study of pertinent cartobibliographies will allow you to determine the availability of maps on a particular geographic area or theme, and their location in one or more libraries.

Cartobibliographies tend to list historical or old maps (usually pre–1900). However, the compilation of maps is not confined to historical topics; on the contrary, much beginning cartographic research focuses on current topics. Consequently, there is a need for current base maps on which these compilations can be made.

One problem with current research themes is copyright. Most maps published in the United States by private or commercial companies and individuals are protected by copyright laws for a specified number of years (see *Copyright Basics*, Circular R1 [Washington: Library of Congress, Copyright Office, 1984]). Therefore, using such secondary compilations for base maps is not recommended unless permission is obtained from the copyright owner. Consequently, it is better to use maps that are not covered by copyright.

A major source of uncopyrighted material in the United States is maps produced by the federal government. Specifically, the topographic maps produced by the U.S. Geological Survey are not copyrighted and are often used for base data information.

Locating Base Maps

Equally important are base maps for compiling thematic maps. Several collections of base or outline maps have been published to aid in the construction of thematic maps of the United States, particularly choropleth maps based on

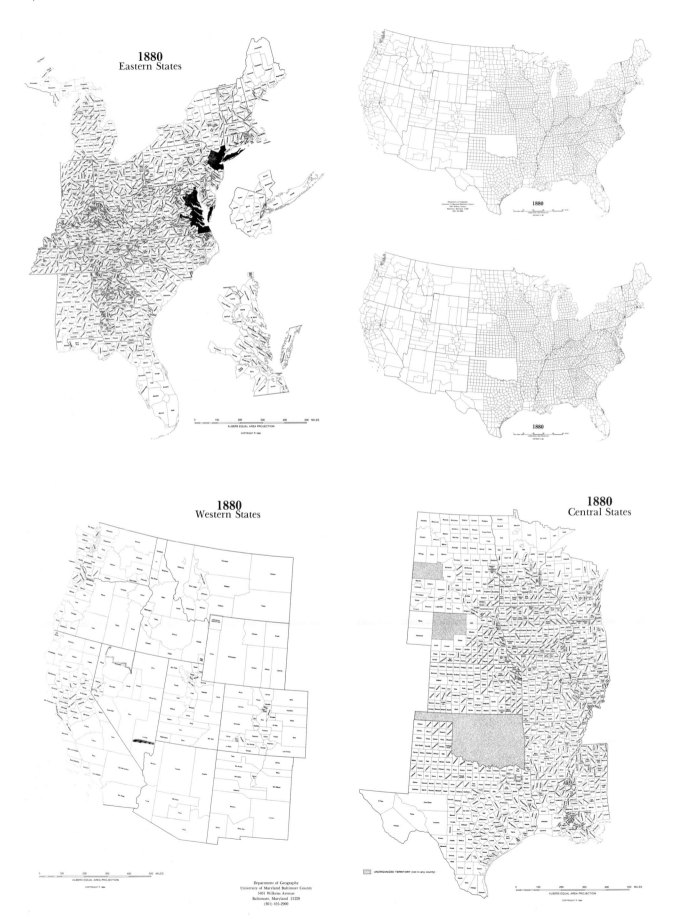

1880
Eastern States

1880

1880

1880
Western States

1880
Central States

ALBERS EQUAL AREA PROJECTION

Department of Geography
University of Maryland Baltimore County
3401 Wilkens Avenue
Baltimore, Maryland 21228
(501) 455-2900

UNORGANIZED TERRITORY (not in any county)

FIGURE 1.19
County outline maps. A combination of four plates illustrating the types of maps contained in the atlas *Historical U.S. County Outline Map Collection, 1840–1980* (Rabenhorst and Earle 1984). The original format of each plate is 17″ × 23″. This atlas may be purchased from the Department of Geography, University of Maryland Baltimore County, Catonsville, MD 21228. (University of Maryland Baltimore County)

state and county subdivision. Of special interest is a series of outline maps of the United States prepared by Thomas D. Rabenhorst and Carville V. Earle under the title, *Historical U.S. County Outline Map Collection, 1840–1980* (Baltimore: University of Maryland Baltimore County, Department of Geography, 1984)—see figure 1.19. This series contains maps of the United States showing the county boundaries at the time of each decennial census to correspond with the data collected by the U.S. Bureau of the Census. This map series was derived from maps issued by the U.S. Department of Agriculture in the 1940s (these maps may be found in many government depository libraries).

The U.S. Bureau of the Census has issued individual state maps showing boundaries of counties and minor civil divisions for each decennial census year from 1940 to present. The 1970 maps have been reproduced in John L. Andriot's *Township Atlas of the United States* (McLean, VA: Andriot Associates, 1979).

Historical research in the southeastern United States will be facilitated by a similar se-

ries of county outline maps compiled by Stephen S. Birdsall and John W. Florin, *A Series of County Outline Maps of the Southeastern United States for the Period 1790–1860*, Map Study No. 2 (Chapel Hill: University of North Carolina, Department of Geography, 1973).

Although no similar series of maps exists for the northeastern United States before 1840, the Newberry Library (a Chicago library specializing in pre–1900 American history) has compiled a data base of historical state and county boundary-line changes since 1788 for 14 northeastern and midwestern states: John H. Long, ed., *Historical Atlas and Chronology of County Boundaries, 1788–1980*, 5 vols. (Boston: G.K. Hall, 1984). This publication covers the states of Pennsylvania, Maryland, New Jersey, Delaware, Ohio, Indiana, Illinois, Michigan, Wisconsin, Missouri, Iowa, Minnesota, South Dakota, and North Dakota.

A listing of other historical atlases, many of which show historical boundary changes, is found in Chapter 5, "Historical Atlases," pages 53–67, in Ronald E. Grim, *Historical Geography of the United States: A Guide to Information Sources* (Detroit: Gale Research, 1982).

The references mentioned in this bibliographical essay are not meant to be definitive, and are only the beginning. They are the key sources to be used in gathering data for map compilation. As your research progresses on a particular topic, other sources of cartographic data or base maps will become apparent and should be utilized.

SELECTED READINGS

American Society of Photogrammetry. 1960. "Procurement of Photography." Chap. 2 in *Manual of photographic interpretation*. Washington.

Chaplin, Edward L. The value of cartobibliographies and the technique of their compilation. *Surveying and Mapping* 20: 76–84.

Colwell, R.N., ed. 1983. *Manual of remote sensing*. Vol. I, *Interpretation and applications*. 2d ed. Washington: American Society of Photogrammetry.

Curran, Paul J. 1985. "Aerial Photography." Chap. 3 in *Principles of remote sensing*. New York: Longman.

Ehrenberg, Ralph. 1982. *Maps and architectural drawings*. Chicago: Society of American Archivists.

Gorman, Michael, and Paul W. Winkler. 1978. *Anglo-American cataloging rules*. 2d ed. Chicago: American Library Association. (See p. 83–109.) (See also Stibbe 1982.)

Larsgaard, Mary L. 1978. *Map librarianship: An introduction*. Littleton, CO: Libraries Unlimited, Inc.

Larsgaard, Mary L. 1984. *Topographic mapping of the Americas, Australia, and New Zealand*. Littleton, CO: Libraries Unlimited, Inc.

Lillesand, Thomas, and Ralph W. Kiefer. 1987. Chap. 9 in *Remote sensing and image interpretation*. 2d ed. New York: John Wiley & Sons.

Makower, Joel, ed. 1986. *The map catalog*. New York: Vintage Books-Tilden Press.

McDermott, Paul D., and Ronald C. Grim. 1976. Maps of the Mullan Road. In *Proceedings of the American Congress of Surveying and Mapping 36th Annual Meeting*, 212–226.

Modelski, Andrew M., comp. 1975. *Railroad maps of the United States*. Washington: Library of Congress. (See p. 75.)

Muehrcke, Phillip C. 1978. *Map use: Reading, analysis, and interpretation*. Madison, WI: JP Publications. (See p. 359–384.)

Nichols, Harold. 1982. *Map librarianship*. 2d ed. London: Clive Bingley. (See p. 120–180.)

Shupe, Barbara, and Colette O'Connell. 1983. *Mapping your business*. New York: Special Libraries Association.

Stibbe, Hugo L. P., ed. 1982. *Cartographic materials: Manual of interpretation for AACR2*. Chicago: American Library Association. (See also Gorman and Winkler 1978.)

Thompson, Morris M. 1979. *Maps for America: Cartographic products of the U.S. Geological Survey and others*. Washington: U.S. Government Printing Office.

Verner, Coolie. *"Carto-bibliography,"* Information Bulletin-Western Association of Libraries. vol. 7, p. 31–38.

Winch, Kenneth L. 1976. *International maps and atlases in print*. 2d ed. New York: R.R. Bowker.

SOURCES OF MAPS AND REMOTE-SENSING IMAGERY

The following list contains contact information for some of the major sources of maps and remote-sensing imagery, including aerial photographs. Information is subject to change. Each source includes a brief notation regarding the types of products available.

Organization	Services
1. Agricultural Stabilization and Conservation Service Aerial Photography Field Office P.O. Box 30010 Salt Lake City, UT 84130-0010 801/524-5050	Aerial photos.
2. Defense Mapping Agency DMA Combat Support Attn: DOA Washington, DC 20315-0010	Limited map distribution to public. Primarily general-purpose and nautical charts of foreign areas.
3. EOSAT 4300 Forbes Blvd. Lanham, MD 20706 800/344-9933 In Maryland: 301/552-0500	(Earth Satellite Corp.) LANDSAT remote-sensing imagery. (See "Ordering LANDSAT and SPOT Imagery" in this section.)
4. Library of Congress Geography and Map Div. Washington, DC 20540	World's largest map collection. Assorted maps; good source for foreign maps. (See "Library of Congress" and "Cameo" in this section.)

Organization	Services
5. National Archives and Records Service Cartographic Archives Div. Washington, DC 20408 703/756-6700	Historical maps; aerial photos created by U.S. government agencies. (See "National Archives" and "Cameo" in this section.)
6. National Geographic Society 17th and M Streets, NW Washington, DC 20013 202/921-1200	General-purpose maps of the entire world, regions, nations, and cities; special thematic maps. Contact for list and prices.
7. NOAA/NOS Distribution Branch (N-CG 33) 6501 Lafayette Riverdale, MD 20737 301/436-6990	(National Oceanographic and Atmospheric Administration/National Ocean Survey.) Aeronautical and nautical charts.
8. SPOT Image Corp. 1897 Preston White Drive Reston, VA 22091–4326 703/620-2200	SPOT-satellite remote-sensing imagery. (See "Ordering LANDSAT and SPOT Imagery" in this section.)
9. USDA/Soil Conservation Service Information Division P.O. Box 2890 Washington, DC 20013 202/447-4543	(U.S. Department of Agriculture/Soil Conservation Service.) Aerial photos, soil surveys.
10. U.S. Geological Survey Aerial Photography Field Office Federal Center Building 25 Denver, CO 80225 303/236-7475	Aerial photos.
11. U.S. Geological Survey Denver Distribution Branch Box 25286 Federal Center Denver, CO 80225	Topographic maps—*Western* United States.
12. U.S. Geological Survey NCIC 507 National Center Reston, VA 22092 703/860-6045 800/USA-MAPS	(National Cartographic Information Center.) Information service for maps and aerial photography. (See "NCIC" in this section.)

Organization	*Services*
13. U.S. Geological Survey Washington Distribution Branch 1200 Eads Street Arlington, VA 22202 202/343-8073	Topographic maps—*Eastern* United States.
14. U.S. Government Printing Office Superintendent of Documents North Capitol and H Streets, NW Washington, DC 20402	Central Intelligence Agency and miscellaneous maps.

PROJECTS

The projects for this section provide experience in determining map coverage for project areas, annotating the content of maps for bibliographic purposes, and determining aerial and satellite imagery coverage for a given region.

PROJECT 1A

IDENTIFYING SOURCE MAPS

Objective

This project provides you with experience in determining existing map coverage for a given area. The geographic area we have selected is the Midwest, encompassing portions of Illinois and Missouri. You are to determine what USGS maps are available for use in developing new cartographic products for this project area.

To delineate map coverage for a study area, start with the index maps. Index maps allow you to determine what maps exist in a series, at a given scale, for a predetermined area. In this case, we have extracted appropriate index maps from USGS catalogs and reprinted them here as figures 1.8, 1.20, 1.21, 1.22, and 1.23.

Figure 1.20 shows the index to the International Map of the World Series at 1:1,000,000 scale. Figure 1.21 is the index map for the State Map Series at 1:1,000,000 and 1:500,000 scales. Figure 1.22 displays the 1 × 2-Degree Series at 1:250,000, and figure 1.23 shows the large 1:100,000-scale 30 × 60-Minute Series. Finally, the largest-scale map series used in this exercise, the 7.5-minute Topographic Map Series at 1:24,000, is depicted in figure 1.8.

FIGURE 1.20
USGS Index—International Map of the World Series, scale 1:1,000,000. (USGS)

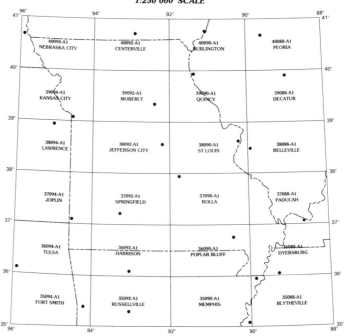

FIGURE 1.22
USGS Index—1 × 2-Degree Map Series, scale 1:250,000. (USGS)

FIGURE 1.21
USGS Index—State Map Series, scales 1:500,000 and 1:1,000,000. (USGS)

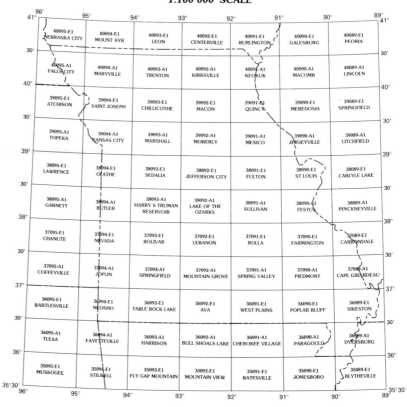

FIGURE 1.23
USGS Index—30 × 60-Minute Map Series, scale 1:100,000. (USGS)

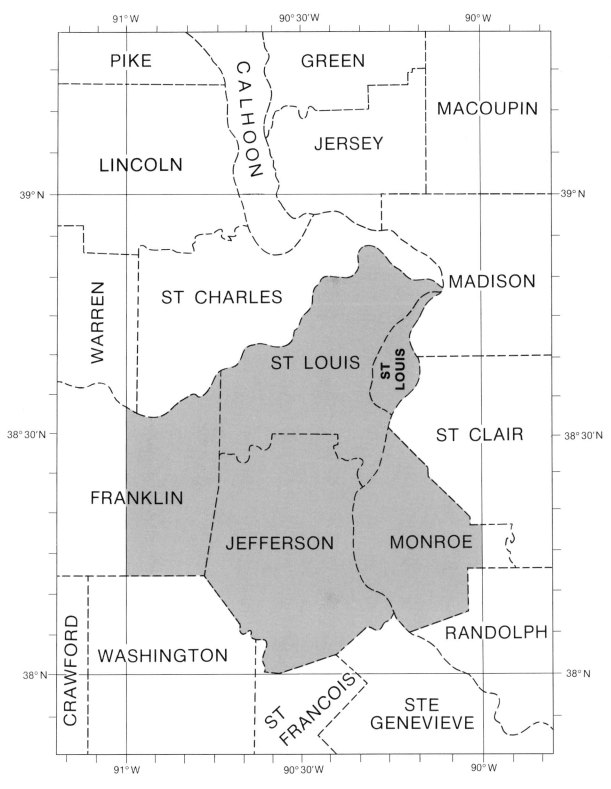

FIGURE 1.24
Study area—Missouri and Illinois.

MAP INVENTORY SHEET

	SHEET NAME		SHEET NAME
IMW Series 1:1,000,000			
SM Series 1:500,000			
1 x 2 - Degree 1:250,000			
30 x 60 - Minute 1:100,000		**7.5 - Minute 1:24,000**	
7.5 - Minute 1:24,000			

FIGURE 1.25

Map inventory sheet. The sheet is divided into sections within which you record the name of every map that pertains to part or all of the study area. An inventory sheet should be compiled whenever an investigation is undertaken.

The display format of the index maps is a new design being implemented by USGS. State indexes are being published in booklet form, with complete coverage planned for all 50 states and possessions of the United States. For each geographical entity covered, one booklet will feature index maps, and another will provide a list of available maps, their prices, information for ordering maps, and locations where the maps may be examined.

Materials and Equipment

1. Set of colored pencils
2. Drafting pencil
3. Materials included in this section
4. Figures 1.8, 1.20, 1.21, 1.22, 1.23

Procedure

1. To determine map coverage for a geographical entity to be studied and mapped, first delineate the study area. The study area for this exercise is shaded in figure 1.24. It includes the city of St. Louis, St. Louis County, part of Franklin County, Jefferson County, and part of Monroe County.

2. Examine each index map sequentially. Compare the study area (figure 1.24) to the corresponding location on the map indexes. Note the name of every map available in each series that covers (in part or total) that location. Write the map names in the spaces provided on the Map Inventory Sheet (figure 1.25).

3. Using colored pencils, outline on the map in figure 1.24 the area covered by the maps in each series. It is not necessary to map the coverage of the state map series. Use a different colored pencil for the outline of sheets drawn at each scale—for example, yellow for 1:24,000 sheets, blue for 1:100,000, red for 1:250,000, and green for 1:1,000,000. The outlines of these sheets can be coordinated to the grid information provided around the neatline of figure 1.8.

PROJECT 1B

CATALOGING A MAP

Objective

One task frequently encountered by cartographers is that of map cataloging. With specialized training, it is possible for the cartographer to become a map librarian. Since many libraries and mapping agencies maintain map collections, the cartographer or map librarian must catalog maps in a manner consistent with the needs of users.

Map cataloging involves three separate steps. First, the content of each map is examined and analyzed. Next, a complete description of the map is developed, including all relevant details. Usually this information is extracted from the sheets and is written onto standardized forms using subject fields to ensure consistency in the annotation process. Finally, the data is entered into a computerized data base. From this data base, researchers can determine the availability of different products by using a computer terminal to conduct the search.

Sometimes you will undertake the analysis of a map collection for purposes other than development of a collection data base. Because many collections have never been described systematically, you may wish to compile a cartobibliography describing that collection. A complete annotation of a map in the Library of Congress collection is shown in figure 1.26.

You also may wish to describe all maps relevant to a special subject of interest, such as railroads, immigrant wagon trails/roads, or land ownership. Many collections of this type have a distinct historical focus. Their function is to illustrate for other researchers the maps available in different collections with regard to content, specific scale, and the originating agency or author. The sample in figure 1.27 illustrates the format of a typical annotation in a collection describing railroad maps.

Note that in an annotation, the amount of detail for a given map is somewhat less than that required for cataloging the map for a data base.

Map cataloging is a difficult task. Typically, a librarian may take several hours to completely annotate a product. Part of the difficulty in annotation is attributable to the fact that the content of the map may be expressed in different symbolization and languages, some of which may be unfamiliar. Other problems result from the varied forms, sizes, and scales at which maps are drawn and reproduced. Because of these difficulties, librarians often use cataloging guides which describe how the product should be analyzed and described. An example of one guide written for this purpose is *Anglo-American Cataloging Rules*, second edition, or AACR2 (see "Materials and Equipment" for this project).

This project involves map description and annotation. The map illustrated in figure 1.28 demonstrates the type of content which must be examined by the cataloger. The major components of the map have been enlarged so you can see pertinent details. The map used here, "The Southwest," was reproduced in color as a supplement in the *National Geographic Magazine*, November, 1982. In this exercise you are to annotate this map.

```
rist washington;file=maps          10 OF 11 RECORDS
08/01/85              [MAPS]            [FIND]    [MUMS]       PAGE   1 OF  1
0*UPD* DISPLAYED RECORD HAS BEEN VERIFIED.                                  112
VERIFIED     MAP RECORD                              AACR 2
```

```
001  85-693161 MAPS
050  G3851.E635 1984 .N3
052  3851
110  National Geographic Society (U.S.)  Cartographic Division.
245  Tourist Washington / produced by the Cartographic Division, National
     Geographic Society ; John B. Garver, Jr., chief cartographer ; William
     M. Palmstrom, map design ; drawn by John G. Weber ; compiled by Harold
     A. Hanson, National Geographic Cartographic Division. Metropolitan
     Washington / drawn by James E. McClelland ; compiled by David B.
     Miller, National Geographic Cartographic Division.
255  Scale [ca. 1:13,500].
255  Scale [ca. 1:110,000].
034  a 13500
034  a 110000
260  Washington, D.C. : National Geographic Society, c1984.
265  National Geographic Society, 17th & M Sts. N.W., Washington, DC 20036
300  2 maps on 1 sheet : both sides, col., plastic-coated paper ; each 26 x
     48 cm., sheet 28 x 50 cm., folded to 14 x 9 cm.
500   August 1984.
500  Includes notes and indexes to points of interest.
500  Metropolitan area insets: 1800: 14,000 people -- 1846: 50,000 people --
     1984: 633,000 people, 3.3 million in metropolitan area.
651  Washington (D.C.)--Maps, Tourist.
651  Washington Metropolitan Area--Maps, Tourist.
700  Hanson, Harold A.
700  Garver, John B.
700  Palmstrom, William N.
700  Weber, John G.
700  McClelland, James E.
700  Miller, David B. (David Byers)
740  Metropolitan Washington.
017  VA 173-014 U.S. Copyright Office
007  aj-canzn
039  2 3 3 3
```

FIGURE 1.26

Map annotation printout. This example demonstrates the type of information obtainable using the map data base compiled by the Library of Congress. The map being described here is "Tourist Washington," a National Geographic Society production. (Library of Congress)

PROJECT 1C

DETERMINING AVAILABILITY OF AERIAL PHOTOGRAPHY

(Part 1)

Objective

Using the APSRS sheet (figure 1.10) and the APSRS code explanations (figure 1.11), write a report discussing the availability of photography for the geographic location lat. 59°07' N, long. 135°37' W. The photos will be used in a historical study of timbering operations of the area. You are working with a limited budget, and therefore you can purchase no more than 10 pieces of imagery. Determine the availability of coverage and recommend specific imagery to be acquired. In your report you must justify each piece of imagery that you recommend.

Materials and Equipment

1. APSRS sheet (figure 1.10)
2. Code explanation (figure 1.11)
3. Pencil and notebook paper
4. Yellow marker pen (highlighter)
5. World atlas

Procedure

1. Using an atlas, identify the location of lat. 59°07' N, long. 135°37' W and indicate the closest town.

2. Review the APSRS sheet (figure 1.10). Identify all imagery that falls within the 7.5' quadrangle sheet whose SE corner is at lat. 59°07' N, long. 135°37' W. Highlight these entries with the yellow marker.

3. Consult the code explanation sheet (figure 1.11) to determine which photos have the desired qualities you are looking for. Select up to 10 pieces of imagery that will meet the criteria outlined in the project objective.

4. Write a report recommending your choice of imagery and the rationale for each selection.

(Part 2)

Using a set of geographic coordinates of your choice, write or call NCIC (see "Sources") and request an APSRS sheet for that location. From the information they provide, order an aerial photograph for your personal use. Most photographs cost between $5 and $10, depending on source.

PROJECT 1D

IMAGE LOCATION: PATH AND ROW INDEXING

Objective

This exercise will provide experience in locating geographic positions on the WRS Index Sheets (LANDSAT World Reference System), and in determining the row and path numbers necessary for ordering imagery.

Materials and Equipment

1. Pencil
2. Straight edge
3. X-ACTO knife
4. Figure 1.30 (WRS Template)
5. Figure 1.31 (WRS Index Map)

Procedure

1. On the section of WRS Index Map in figure 1.31, find the position of the following pairs of geographical coordinates: (a) 33°10′ N, 96°50′ W, (b) 22°15′ N, 97°50′ W, (c) 39°00′ N, 94°30′ W.

2. Record in the space provided the names of the cities at these three positions:
   (a) 33°10′ N, 96°50′ W: City _____ Path # _____ Row # _____
   (b) 22°15′ N, 97°50′ W: City _____ Path # _____ Row # _____
   (c) 39°00′ N, 94°30′ W: City _____ Path # _____ Row # _____

3. Determine the row and path positions to be used in the procurement of LANDSAT images of the three areas and enter them in the space provided. The satellite crosses the Earth's surface in a northeast-to-southwest course. Row lines tend to follow parallels of latitude on the index sheet. The numbers identifying paths increase from east to west, and the rows increase from north to south. The center of an image is at the intersection of a path and row.

4. Determine the area covered by each image. Place a template of the typical areal coverage of an image over the intersection point, and trace the outline of the coverage. To make a template, cut out figure 1.30 (or photocopy it at 100% size). Using an X-ACTO knife, carefully cut out the rhomboid shape identified as the Normal Scene Area. This shape conforms to the oblique path of the satellite over the Earth.

5. Determine the row and path numbers of all imagery necessary to completely cover the study area outlined on the index map in figure 1.31.

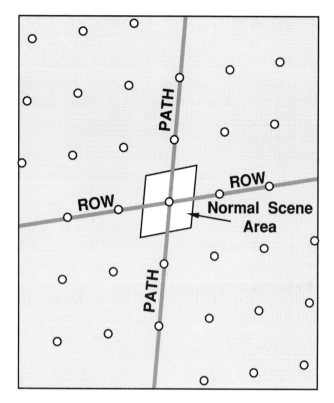

FIGURE 1.30
World Reference System location template. This template is used to determine the areal coverage of a LANDSAT image. It is centered over the intersection of a path and row. (USGS)

FIGURE 1.31
Section of LANDSAT index map. To determine the availability of satellite imagery, you must relate the desired geographic location to a particular row and path number, indicating the point at which an image was taken. (USGS)

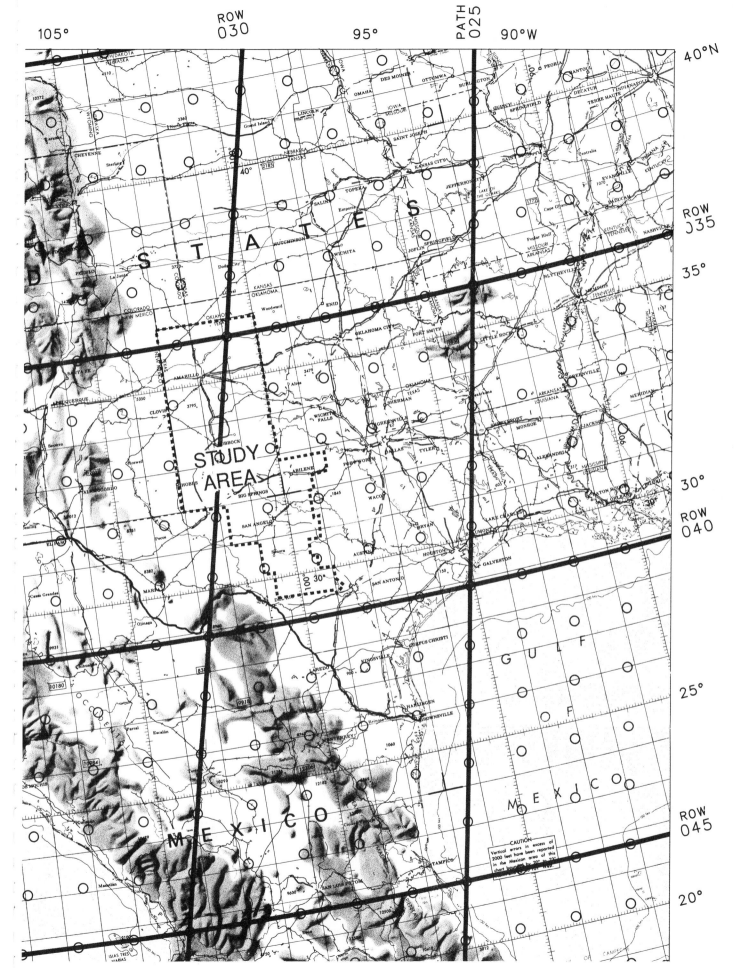

PROJECT 1E   **LANDSAT IMAGE ANNOTATION**

Objective

This project provides you with experience in deciphering the annotation strip at the bottom of LANDSAT imagery.

Materials and Equipment

1. Annotation strip in figure 1.32
2. Refer to "Annotation of LANDSAT Imagery" earlier in this section.

Procedure

In this project you will interpret the LANDSAT annotation strip in figure 1.32. Complete the following form by first listing the annotation asked for, and then verbally describing its meaning.

Image Date:

Image Center:

Path Identifier:

Nadir:

Sensors and Spectral Bands:

Sun Angle:

Geometric Correction, Processing, and Transmission:

Agency and Project:

Image Identification Number:

14FEB79 C N20-11/W104-02 D046-030 N N19-59/W103-58 M 5 D
SUN EL37 A098 R2U-ND-N H2 NASA LANDSAT E-2 037-12524

FIGURE 1.32
LANDSAT annotation for Project 1E.

2

Field Mapping

INTRODUCTION

In this section we will present various field-mapping techniques. While it is not possible to cover all areas of field mapping, some of the simpler methods will be discussed. You will not become a certified surveyor by studying this section; in fact, formal surveying will not be addressed. But, we hope that a basic knowledge of field-mapping techniques will be acquired.

Although the world's inventory of accurate maps grows daily, you will not always be able to find maps with the desired scale and detail for your needs. This is particularly true when very large-scale maps are required. For this reason it is advantageous for you to know how to go into the field and prepare a map of sufficient detail and accuracy to successfully complete a project.

The ability to measure distances and angles precisely is paramount to constructing a realistic representation of an area. These measurements can be obtained in a variety of ways which yield results of varying levels of accuracy. Three such methods are discussed here: *foot-made* maps, *compass-traverse* maps, and *plane-table* maps.

FOOT-MADE MAPS

One advantage of foot-made maps is that they require no special equipment. The key to successful foot-made maps lies in your ability to accurately pace off distances. This requires practice. To determine the length of your pace requires setting up a test course. To do this, measure a distance of 100 feet between two points on level ground. Practice pacing off the distance by walking between the two points. Start with your left foot and count each time your right foot strikes the ground—this is a stride (two paces). Determine how many strides you take to cover the 100-foot distance. Do this a number of times in each direction, practicing until the number of strides you take to cover the distance is very consistent. If you consistently take 20 strides to cover the 100-foot distance, then your stride is 5 feet ($100 \div 20 = 5$).

Once you have mastered your stride, you can put it to work in the field together with a principle learned in geometry: if the lengths of three sides of a triangle are known, the angles can be computed. By striding off distances between known points and forming triangles (a process known as triangulation), you can create

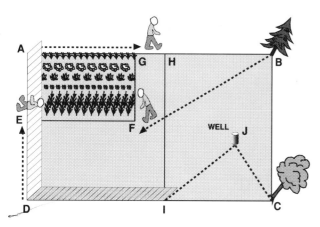

FIGURE 2.1
Pacing off a foot-made map. Begin by pacing off part of the perimeter (**D–A** and **A–B**) and then complete a triangle by pacing from **B** back to **D**. Repeat a similar procedure to establish the dimensions of triangle **BCD**. Now that the perimeter's size and shape have been determined, you can begin to locate interior features. The position of the garden and well can be paced off, as can the line **H–I**.

a map that accurately ties together the desired features.

For example, to map the features illustrated in figure 2.1, begin by establishing the outline of the area to be mapped—a rectangle including a garden and trees B and C. Stride off the distances from D to A, A to B, and then B to D. Now you know the lengths of the three sides of the triangle, and you can draw them on a field-sketched map. Similarly, you can pace off triangle BCD, and draw it to complete rectangle ABCD. With the outer boundaries of the area completed, the interior details can be added using the same triangulation method (lines EF and FG, which define the other two sides of the garden, and HI). Likewise, the well at point J can be located by forming a triangle from point J and two other points, such as I and C.

There is potential for some error due to variations in terrain (strides tend to vary when walking up and down hills). This can be minimized by knowing your stride measurements and how they change with different slopes. Despite this potential for error, the method is a quick and reasonably accurate way of field mapping, and it requires no special equipment.

Once the field sketch is completed and you are satisfied that your notes and measurements are accurate, you can redraw the map using a straightedge, scale, and bow compass. If significant discrepancies appear, you may have to return to the field to double-check your measurements.

COMPASS-TRAVERSE MAPS

The value of the compass-traverse map over the foot-made map is that it requires considerably less walking, thereby reducing the amount of time required to make each traverse. (A "traverse" is a sequence of lines connecting a series of points across a plot of ground.)

While compasses vary in sophistication, all of them operate on the same principle: the compass needle always points in the direction of magnetic north. To establish a line on a compass-traverse map it is necessary to know (1) the origin, (2) the length, and (3) the azimuth of the line. Azimuth is read clockwise in degrees, beginning with 0° at north, 90° east, 180° south, and 270° west (figure 2.2).

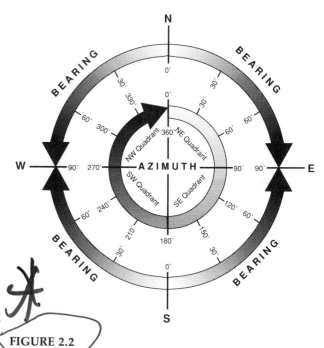

FIGURE 2.2
Compass bearing and azimuth. *Azimuth* is measured from north in a clockwise direction with east at 90°, south at 180°, west at 270°, and north at 0° or 360°. *Bearing* is measured from north, in an east or west direction, increasing to a maximum of 90°; and from south, in an east or west direction, increasing to a maximum of 90°. An *azimuth* of 120° (denoted as **Z 120°**) would have a *bearing* of south 60° east and be denoted as **S 60° E**.

FIGURE 2.3
Magnetic declination. Compare a true-north line, such as the boundary between North Dakota and Montana, to the magnetic-declination line extending from western North Dakota to Arizona. Magnetic north is about 14° east of true north.

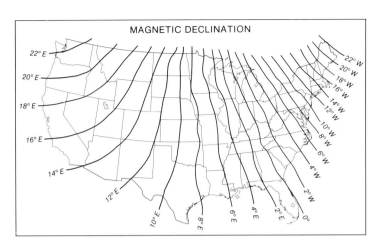

Note that there is a difference between *magnetic north* and *true north*. While true north remains constant regardless of your position on the Earth, magnetic north varies in relation to true north. This results from the magnetic pole of the Earth not coinciding directly with the North Pole. Therefore, depending upon your location, a compass needle may point several degrees away from true north. This variation is referred to as *magnetic declination* (figure 2.3).

Not only does magnetic north vary from place to place, it also slowly changes over time at the same place. Frequently, detailed maps, such as the USGS topographic maps, not only list the magnetic declination for the date the map was produced, but also list the annual change in declination. Though magnetic declination will not affect the relationship of measurements on a traverse, it will affect the overall orientation of the traverse with regard to true north.

A problem likely to have more severe consequences on the accuracy of a traverse is local anomalies. If a compass is brought into close contact with a piece of iron or steel, the needle can be deflected away from magnetic north. You must be aware of these effects, and avoid taking compass bearings when in close proximity to automobiles, buildings with many steel reinforcements, railroad tracks, etc.

There are two categories of compass traverses. The *closed traverse*, which actually encompasses an area so that the beginning and ending points are the same (figure 2.4), and the *open traverse*, which does not have beginning and ending points in common.

Closed Compass Traverse

To begin a closed traverse, establish a point of origin (A in figure 2.4). This should be clearly defined by using a road intersection, a specific utility pole, or some other easily identifiable point. As with the foot-made map, the outer boundary of the area to be mapped should be established first; this provides a framework within which to link the rest of the map detail.

Stand at the point of origin (A in figure 2.4) and face the next point on the boundary to be located (B). Take a compass bearing by holding the compass so that the north-pointing needle is aligned with the 0 on the compass. Now sight along the compass to point B (figure 2.5) and read the compass azimuth (63°). On a sketch map, record the azimuth. Stride off the distance (225 feet, figure 2.4) and record the results on the sketch map.

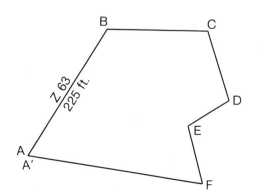

FIGURE 2.4
Closed traverse.

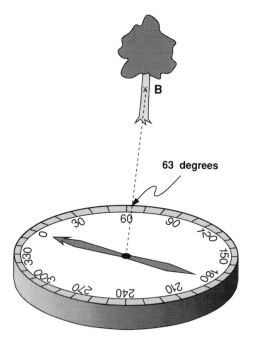

FIGURE 2.5
Sighting with a compass. When sighting with a compass, keep the north end of the compass needle aligned with 0° (north on the compass face). Holding the compass in this position, sight to the desired object and read the azimuth that falls along the line between the object and the center of the compass. In this example the tree is sighted at 63°.

Now you are at point B, and ready to locate point C. Repeat the procedure used to locate point B. Continue locating points and recording the results on the sketch map until arriving at the final point, called A'.

When all measurements and readings have been gathered and recorded correctly, you can return from the field and neatly redraw the traverse. This is accomplished with a scale, straightedge, and protractor. The redrawn traverse is shown in figure 2.6. Only with the greatest fortune will the origin (A) and final point (A') coincide exactly. Almost always it will be necessary to "close" the traverse.

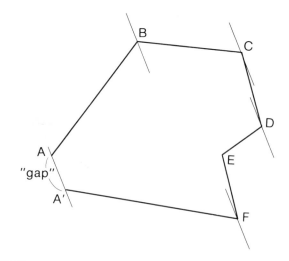

FIGURE 2.6
The space between **A** and **A'** is the "gap" that usually occurs when creating a closed-traverse map.

Closing the Traverse

This procedure assumes that the closure error results from a series of consistent compounding errors, rather than from one major error on one leg of the traverse. Initially, your "gaps" may be fairly large; however, as your level of proficiency increases, your margin of error should diminish. An acceptable error level is 1–3% of the total length of the traverse.

Figures 2.7 and 2.8 illustrate how to close a traverse. Begin by drawing a straight line that is the same length as the traverse (figure 2.7). Place the point of origin (A) on the left and the end point (A') on the right. At the end point (A'), draw a line perpendicular to the traverse line and equal in length to the gap distance (from figure 2.6). Now draw a line from the top of this line, back to the point of origin (A) to form a triangle. Using a scale, mark off (on the base line of the triangle) the distance of each leg of the traverse in order, beginning at the origin (A). Label the points along the traverse base

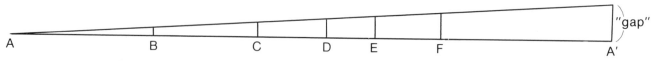

FIGURE 2.7
Determining the amount of correction needed at each point when closing a traverse. The "gap" distance is plotted at one end of a line drawn to scale, which represents the total length of the closed traverse. The lettered points along the traverse line are the scaled locations of the same points along the traverse in figure 2.6.

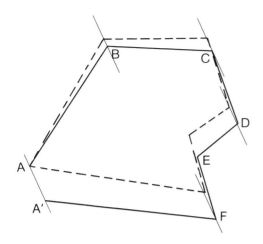

FIGURE 2.8
Closing the traverse. The strike lines drawn through points **B** through **F** on the traverse must be parallel to the line drawn from **A'** to **A**. Then, the "corrected" distances determined in figure 2.7 are marked off at each point. Connecting these new points (dashed lines) will close the traverse.

line (B,C,D,E,F) and draw perpendicular lines from these points upward from the baseline.

On the compass-traverse drawing (figure 2.8), strike a line through each point, parallel to the gap line. Transfer the distances measured from the upper line of the triangle in figure 2.7 to the appropriate parallel strike lines on the compass-traverse drawing (figure 2.8). Now, based on these new measurements, redraw the traverse to close the gap (dashed lines in figure 2.8).

Filling in the Detail

Generally, the easiest way to fill in the detail of a compass traverse is to use the triangulation method. Sight from known points on the established boundary to points within the interior. Figure 2.9 shows how point G can be located by sighting from points B and C. As an additional check, a third point can be used (point E, for example). By using the triangulation method, it is possible to locate points using only azimuth readings, and thereby eliminate the need for pacing or other forms of measurement.

Open Compass Traverse

The open compass traverse is used for field inventory and route exploration, rather than the more conventional area map discussed above.

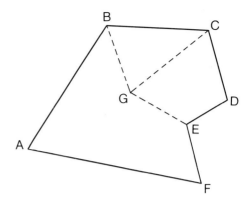

FIGURE 2.9
Additional points can be added within a closed traverse using the method of triangulation.

When constructing an open traverse, there is no way to check for or correct errors, such as the "closure" check of the closed traverse. However, this is not to say that open traverses are necessarily less accurate. Accuracy depends on you and your skills in reading a compass and pacing off distance.

In general, the open traverse is used for such tasks as mapping the course of a river, or for inventorying distributions such as vegetation, historical structures, or other cultural or physical features. Regardless of subject, the same basic principles hold true: azimuth and distance measurements are used to establish the location of the traverse base line. Then, various features along the base line can be plotted, using the triangulation method (figure 2.10). Here the course of a river has been plotted, and the position of a pond and the corner of a building have been triangulated.

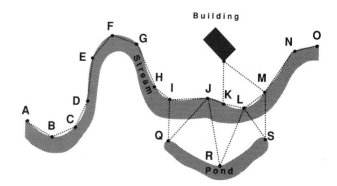

FIGURE 2.10
An open traverse (**A** through **O**) marking the edge of a stream. Additional points (**Q, R, S,** and a building corner) are located using the method of triangulation.

PLANE-TABLE MAPS

Compared to foot-made maps or compass-traverse maps, the plane-table method of mapping is faster and more accurate. For this reason, it has been a favorite form of mapping over the centuries, particularly as the United States was opened for settlement.

Plane-table mapping is an easy technique, employing many of the skills already discussed. While equipment of varying sophistication is available, we will assume a very simple setup in this presentation. This includes a plane table and tripod, sight rule, bubble level, magnetic compass, paper and pencil, and a pocket knife for sharpening pencils in the field.

Once the area to be mapped is determined, erect the plane table at a known point. This may be a stake in the ground, a stump, a fencepost, etc. Using a bubble level, level the plane table surface in all directions and lock it into place. Attach a piece of paper to the surface of the plane table, being careful not to disturb its leveled position. Using a magnetic compass for alignment, strike a north-south line on one side of the paper and label the north end of the line.

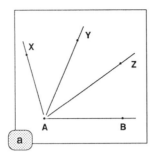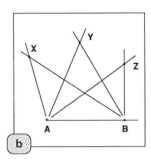

FIGURE 2.12

(a) Sighting from point **A** to points **X**, **Y**, and **Z**. (b) Triangulating from point **B** to points **X**, **Y**, and **Z**. The result is an accurate positioning of the points **X**, **Y**, and **Z** with respect to points **A** and **B**.

Mark a point of origin (A) on the paper to represent the position of the plane table with regard to the area being mapped. You will usually place the origin at the center of the plane table if you are making an area map (closed traverse). However, if you are making an open traverse or linear survey, the starting point is generally positioned to one side of the sheet.

Now, select the first point or object to be located, and align the sight rule so the edge of the rule passes through the point of origin (A) marked on the paper and is directly in line with the sighted point (figure 2.11). Using the edge of the sight rule, strike a line outward from near the point of origin (A) to the point to be located (point X in figure 2.12a). (It is best not to let the line extend to the point of origin, as a number of lines will eventually obliterate the point.) Continue by sighting to other points (Y and Z) and striking lines accordingly onto the drafting surface. Take care not to move or bump the plane table during this process, because errors in alignment will occur.

After sighting to all desired points from the point of origin, construct a *base line*. Establish the base line by sighting to a point (B in figure 2.12b) that is in clear view of the other sighted points (X, Y, and Z). In this example, point B is 200 feet from point A. (The distance between points A and B can be measured with a steel tape.) If the scale of the map is to be one inch = 50 feet, and the actual distance on the ground between points A and B is 200 feet, then the distance on the map between these points will be 4 inches. Use an engineer's scale

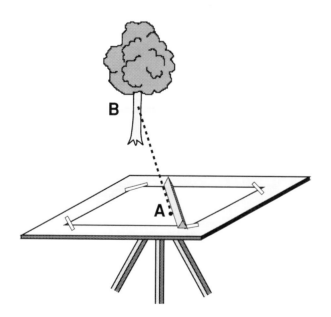

FIGURE 2.11

Sighting along a straight edge on top of a plane table can provide amazingly accurate results if care is taken when aligning the object.

to mark off the appropriate distance on the map. Then, move the plane table to the other end of the base line (point B).

After setting up and leveling the plane table at point B, you must back-sight to the first point (A) to align the drawing. In the example, this is done by slowly rotating the plane table until line AB on the plane table points directly towards point A in the field.

Once you have repositioned and aligned the plane table, points X, Y, and Z can be located by sighting and striking lines to them in a similar manner as was done at point A. The intersection of these pairs of strike lines will accurately locate the points. Continue this process to locate other points, and move the plane table to other locations to expand the area of the map. In this way, you create a map containing the accurate location of many features.

Plane-table mapping can be applied to the open-traverse method to create strip maps of trails, river courses, railroad routes, etc. Project 2E further illustrates the procedures for plane-table mapping.

SELECTED READINGS

Bies, John D., and Robert A. Long. 1983. *Mapping and topographic drafting*. Cincinnati: South-Western Publishing Co. (See p. 107–139, 140–155.)

Blandford, Percy W. 1984. *Maps & compasses: A user's handbook*. Blue Ridge Summit, PA: TAB Books. (See p. 147–174.)

Commission on College Geography. 1968. *Field training in geography*. Technical Paper no. 1. Washington: Association of American Geographers.

Garnier, B.J. 1963. *Practical work in geography*. London: E. Arnold. (See p. 39–61, 62–76.)

Greenhold, David. 1964. *Mapping*. Chicago: The Univ. of Chicago Press. (See p. 203–239.)

Lounsbury, J.F., and F.T. Aldrich. 1986. *Introduction to geographic field methods and techniques*. 2d ed. Columbus, OH: Merrill Publishing Co. (See p. 46–48, 50–66, 75–83.)

Low, Julian W. 1952. *Plane table mapping*. New York: Harper & Row.

Stoddard, R.H. 1982. *Field techniques and research methods in geography*. Dubuque, IA: Kendall/Hunt Publishing Co.

PROJECTS

The projects for this section are designed to familiarize you with the techniques necessary to produce field maps. Projects include foot-made maps, compass-traverse maps, and plane-table maps constructed from field observations and reconstructed from field notes.

PROJECT 2A

FOOT-MADE MAPS

Objective

Foot-made maps are a quick and easy way to gather data for compiling a simple field map. This exercise guides you through a simple mapping problem using foot-made mapping techniques.

Materials and Equipment

1. Drafting pencil and eraser
2. Drafting paper
3. Engineer's scale
4. Triangle or straight edge
5. Bow compass
6. Clipboard
7. Tape measure (100-foot)
8. Field Notes Recording Sheet (figure 2.13)

Procedure

1. Your instructor or a designated student should position about 10 stakes in an open field, covering an area of 1–3 acres. The stakes should define the perimeter of the area to be mapped, as well as any notable points within it. In lieu of stakes, objects like trees, rocks, corners of buildings, and signs may be selected.

2. Determine the length of your stride. While this may appear to be a simple task, most people need to spend a considerable period of time pacing off a 100-foot distance before they can consistently walk the distance in the same number of strides. Do not proceed until you have mastered your stride.

3. Locate the stakes or objects on the course. Record the measured data on the Field Notes Recording Sheet (figure 2.13). An example of how to record the data is shown in figure 2.14. It is often useful to construct a simple sketch map of the relative position of the key points while the data is being collected. If time permits, pace off each distance two or three times and average them to minimize error in measurement.

4. Reconstruct the map from your field notes. Select a scale that allows the entire map to be drawn on a single sheet of drafting paper. (Note: Your instructor may want to assign the scale to be used.) Begin constructing the map by locating the first point (A) on the drafting paper (figure 2.15a). Draw a straight line outward from point A toward the second point (B). Based on the map scale, place the point of a bow compass on A and open it so as to strike an arc with a radius equal to the distance between points A and B. Point B will be located where the arc crosses the straight line (figure 2.15a).

   Next, triangulate to a point common to both points A and B (in this example, point G). This is accomplished by positioning the point of the compass at point A and striking an arc with a radius equal to the distance from point A to point G (figure 2.15b). Now, place the point of the compass at B and strike another arc with a radius equal to the distance from point B to point G (figure 2.15c). The intersection of the two arcs locates point G. Continue this triangulation procedure until all points have been located.

5. Once all points have been located, you may elect to return to the field to sketch in additional features referenced to the plotted points. In this way, you can create a fairly detailed map.

Field Notes Recording Sheet
Foot-Made Maps

Point to Point	Distance	Notes
to		
to		
to		
to		
to		
to		
to		
to		
to		
to		
to		
to		
to		
to		
to		
to		
to		
to		
to		
to		
to		
to		
to		
to		
to		
to		
to		
to		
to		
to		

FIGURE 2.13
Field Notes Recording Sheet for foot-made maps.

Field Notes Recording Sheet
Foot-Made Maps

Point	to	Point	Distance	Notes	
S	to	C1	2 paces	West to point C1	Corn Patch
C1	to	C2	10 paces	Continue west to point C2	
C2	to	C3	12 paces	North to C3	
C3	to	C4	10 paces	East to C4	
C4	to	C1	12 paces	South to C1 (parallel to side of house)	
* C1	to	C3	16 paces	Check on rectangle shape (diagonal)	
C1	to	V1	6 paces	West to V1	Salad Patch
V1	to	V2	8 paces	South to V2	
V2	to	V3	4 paces	East to goldfish pond	
V3	to	V4	4 paces	NE along edge of goldfish pond	
V4	to	C1	5 paces	North to C1	
* C1	to	V2	10 paces	Check on polygon (diagonal)	
C1	to	B1	14 paces	South to B1 (corner of tool shed)	Bean Patch
B1	to	B2	4 paces	East to B2 along edge of tool shed	
B2	to	V2	6 paces	North to V2	
* B2	to	C2	14½ paces	Check on rectangle (diagonal)	
C1	to	M1	12 paces	East along side of house to M1	Melon Patch
M1	to	M2	12 paces	South to M2	
M2	to	M3	3 paces	West to M2 at end of goldfish pond	
M3	to	V4	12 paces	NW to V4 along edge of goldfish pond	
* V4	to	M1	13 paces	Check on polygon (diagonal)	
M2	to	T1	8 paces	South to T1	Tomato Patch
T1	to	T2	7 paces	West to T2	
T2	to	T3	6 paces	North to T3 at edge of goldfish pond	
T3	to	M3	4½ paces	NE to M3 along edge of goldfish pond	
T3	to	T1	9 paces	Check on polygon (diagonal)	
T2	to	P1	10 paces	West to P1 at side of tool shed	Potato Patch
P1	to	P2	6½ paces	North to corner of tool shed	
P2	to	B2	4 paces	West to B2	
V3	to	P3	5½ paces	South southeast along edge of goldfish pond	
P3	to	T3	5½ paces	Along edge of goldfish pond	
* T3	to	P1	10½ paces	Check on polygon (diagonal)	
* V3	to	B3	7 paces	Check on Polygon (diagonal)	

NOTE: 1 pace = 30 inches
* diagonal checks on polygons

FIGURE 2.14
Completed field notes recording sheet.

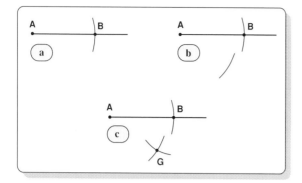

FIGURE 2.15
(a) Using a bow compass to locate triangulated points. Draw a straight line and place point **A** at one end of the line. Open a bow compass to the scaled distance between points **A** and **B**. Place the point of the compass at point **A** and strike an arc that crosses the line. This will locate point **B** in relationship to point **A**. (b) To triangulate to point **G**, set the compass for the distance from point **A** to point **G**. Place the point of the compass at point **A** and strike an arc in the vicinity of point **G**. (c) Repeat this procedure for the distance from point **B** to point **G** by placing the compass point at point **B**. The place where the two arcs intersect is the precise location of point **G** with respect to points **A** and **B**.

PROJECT 2B

RECONSTRUCTING A FOOT-MADE MAP FROM FIELD NOTES

Objective

This exercise gives you experience in reconstructing a foot-made map from field notes. We have specifically designed this project for those times when circumstances prevent you from actually going into the field to record the data.

Materials and Equipment

1. Same as Project 2A Materials and Equipment, 1 through 5
2. Completed Field Notes Recording Sheet (figure 2.14)

Procedure

1. Assume that the person who gathered the data has a stride of 5 feet.

2. When constructing the map, use a scale of 1 inch = 20 feet.

3. Reconstruct the map using the techniques outlined in Project 2A, Procedure, paragraph 4 ("Reconstruct the map from your field notes").

4. Sketch in additional information, based on the recorded field notes.

PROJECT 2C

RECONSTRUCTING AND CLOSING A COMPASS TRAVERSE FROM A FIELD SKETCH

Objective

This exercise gives you experience in reconstructing and closing a compass traverse, based on field-sketch data. This exercise can be used as preliminary experience for Project 2D. It may also serve as a substitute for Project 2D when conditions such as bad weather or lack of time prevent "in the field" data collection.

Materials and Equipment

1. Same as Project 2A Materials and Equipment, 1 through 5
2. Protractor
3. Field sketch map (figure 2.16)

Procedure

1. Using the field sketch (figure 2.16), reconstruct the map perimeter using a scale of 1 inch = 20 feet.

2. Close the traverse following the procedures outlined in this chapter under the heading "Closing the Traverse."

3. Place a piece of tracing paper over the closed traverse and redraw the map perimeter.

4. Fill in the map details. Use triangulation, based on compass bearings obtained from the field sketch.

CLOSED COMPASS TRAVERSE SKETCH MAP

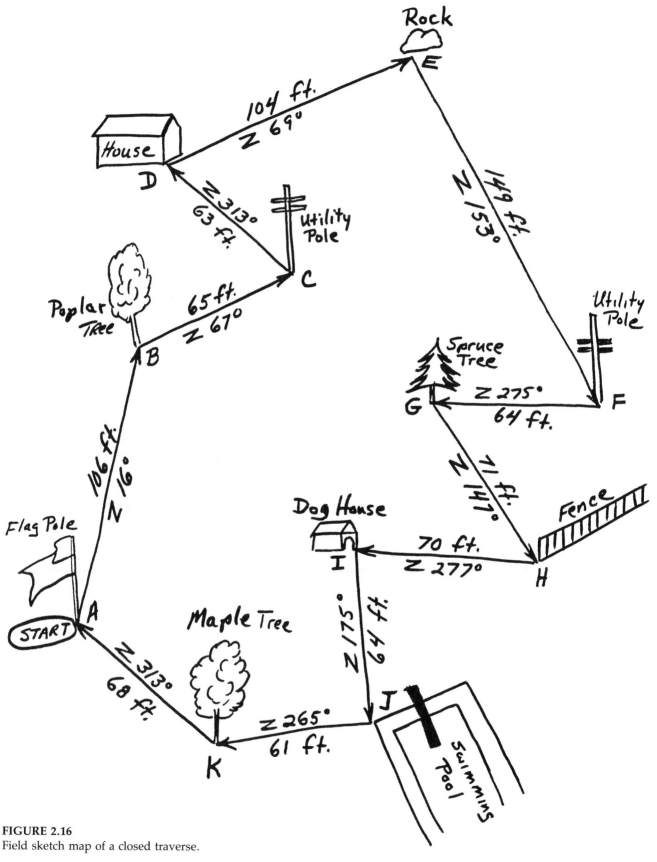

FIGURE 2.16
Field sketch map of a closed traverse.

PROJECT 2D

COLLECTING FIELD DATA FOR A CLOSED COMPASS TRAVERSE

Objective

This exercise familiarizes you with the procedures and techniques necessary for field mapping when using a compass and steel tape. Activities include establishment of the map perimeter, traverse closure, and filling in map detail.

Materials and Equipment

1. Same as Project 2A Materials and Equipment, 1 through 7
2. Protractor
3. Magnetic compass (Brunton)
4. Field Notes Recording Sheet (figure 2.17) (we recommend that you make photocopies for use)

Procedure

1. This exercise is accomplished most easily if students are divided into teams of two or three.

2. Your instructor will establish the perimeter of the area to be mapped. The size of the area should be 1–3 acres. Use various objects such as trees, rocks, corners of wooden buildings, etc. as points to define the perimeter of the map. CAUTION: Avoid objects like automobiles, metal posts, and buildings reinforced with structural steel, as they will cause the compass to deflect and render false bearings. Select one of the objects to be the origin of the traverse.

3. Standing at the point of origin, use the magnetic compass to obtain a bearing to the next point along the map perimeter. The person handling the compass carries a pencil, clipboard, and paper, and records the measurements. Sketch the perimeter as it is being traversed. Place a second point (B) on the sketch and connect it with a straight line drawn to the origin (A). Record the compass bearing along the line drawn between points A and B.

4. The other students in the team are responsible for measuring the distance between points. For best results, use a tape measuring 100 feet or more. For distances longer than the tape, it will be necessary to make the measurement in two or more segments. It is the responsibility of the compass-bearer to keep the person with the tape aligned with the two points being measured. Once the distance has been determined, record it on the sketch map. An example of a sketch map is provided in figure 2.16.

5. Advance to point B and repeat the procedure when sighting to point C. Note: It is often useful to "back-sight" to the previous point to act as a check on the earlier sighting. You will get a reading 180° different if the two bearings are correct.

6. Continue moving to each consecutive point, recording bearings and distances until the traverse is completed by arriving back at the point of origin.

7. Once the field data has been collected, you can accurately redraw the map. Choose a scale that permits the entire traverse to be drawn on one sheet of drafting paper. (Note: Your instructor may select a common scale for the class as a whole.) Most likely, the beginning and end points of the traverse will not coincide upon completion of the initial drafted traverse. This results from errors in bearing and distance measurement. To correct the errors, follow the procedures outlined in this section under the heading "Closing the Traverse."

8. Optional—Your instructor may elect to have you fill in the detail within the map perimeter, creating a detailed map rather than an outline map. Additional points not on the perimeter can be located by triangulation. Much "leg work" can be eliminated by using compass bearings alone to locate these points. While it is most efficient to gather the additional points at the same time the perimeter data is being collected, these points should not be added to the drafted traverse until after the traverse has been closed.

Field Notes Recording Sheet
Compass-Traverse Maps

Point	to	Point	Distance	Bearing	Notes
	to				
	to				
	to				
	to				
	to				
	to				
	to				
	to				
	to				
	to				
	to				
	to				
	to				
	to				
	to				
	to				
	to				
	to				
	to				
	to				
	to				
	to				
	to				
	to				
	to				
	to				
	to				
	to				
	to				

FIGURE 2.17

Field Notes Recording Sheet for compass-traverse maps.

PROJECT 2E   **PLANE-TABLE MAPPING**

Objective

This project acquaints you with elementary plane-table mapping skills. Our intent is to present enough plane-table mapping techniques to enable you to create a reasonably accurate large-scale map. Many of the finer techniques have been omitted, in the interest of time. In fact, to make this exercise workable for as many groups as possible, we have substituted an engineer's scale for the plane table alidade as the sighting device. Therefore, corrections for differences in elevation will not be taken into account in this project.

In this exercise, simple triangulation is used to create either an open traverse or closed traverse. Even with a simple system of triangulation, variations such as the use of quadrilaterals, polygons, and their combinations can be applied. As you become proficient in the simple mapping techniques presented here, you may want to expand your skills in plane-table mapping. An excellent source is *Plane Table Mapping* by Julian W. Low (see "Selected Readings").

A principal difference between plane-table mapping and the other field-mapping techniques mentioned in this unit is that plane-table mapping actually involves drafting the map in the field as data is being gathered. This has the advantage of letting you see what the map looks like as you construct it, and therefore allows you to modify it or add information during the data-gathering process.

As with other projects in this unit, your instructor must designate the area to be mapped. This might be a stream, a trail through a forest, a series of marker posts positioned in a field, or possibly a small, irregularly shaped field. Depending upon the terrain, visibility to successive points, and the distance between the points being mapped, the traverse may range in length from a few hundred feet to nearly a mile. A reasonable expectation for a first attempt at this kind of mapping is a course containing between 10 and 20 points to be mapped.

Materials and Equipment

1. Drafting pencil and eraser
2. Drafting paper
3. Plane table and level
4. Engineer's scale
5. Steel tape measure (at least 100 feet long)
6. Magnetic compass

Procedure

1. Begin by establishing a base line. In Low's book on plane-table mapping, he states:

   "Whenever possible, the base line should be measured on open, flat ground with the two ends so situated that they are intervisible from each other and from other points in the vicinity that will be used as triangulation stations. The location of the base line within the map area will be somewhat dependent on suitable sites; but if the area is large, the base line should be located somewhere in the central part so that the net will not have to be extended too far in any one direction. If the map area is greatly elongated, the base line should be oriented as nearly parallel as possible to the longer dimension of the area.''

Once a suitable place has been selected for the base line, mark each end of it with a flag stake.

2. Accurately measure the length of the base line, using a steel tape measure. If the length of the base line is greater than the tape, take care when measuring to stay directly in line with the two ends of the base line.

3. Using masking tape, mount a piece of drafting paper onto the plane-table surface.

4. Place the plane table at the end point of the base line (point A in figure 2.18a). Level the table in all directions.

5. Using a magnetic compass to determine the direction north, draw a line near the edge of the paper representing north-south alignment, and label the north end.

6. Establish the scale to be used when constructing the map. The scale should be small enough to prevent the map from extending beyond the drafting surface on the plane table. (Note: Your instructor may want to determine the scale to be used.) Place a point (A) on the drafting paper representing the location of the plane table. With the engineer's scale resting on the plane table, sight along its edge to the other end (point B) of the base line. Be sure to keep the edge of the scale on point A when sighting to point B. When alignment is completed, draw a ray from point A outward toward point B. Measure off the length of the base line along this ray, based on the scale selected for the map (figure 2.18a).

7. Keeping the edge of the engineer's scale over point A, sight along it to other triangulation points that must be located. Each time, after alignment, draw a ray outward from point A toward the triangulation point (figure 2.18b).

8. After all triangulation points have been sighted from point A, move the plane table and equipment to the other end of the base line (point B). Level the plane-table surface. Realign the drafted base line with the actual base line by placing the edge of the scale along the drafted base line and then slowly rotating the entire plane-table surface until the edge of the scale is aligned with point A. Lock the plane-table surface in place.

9. Repeat procedure 6, except draw the rays from point B instead of point A (figure 2.18c). The triangulation points are located where the rays from B intersect their counterparts radiating from point A.

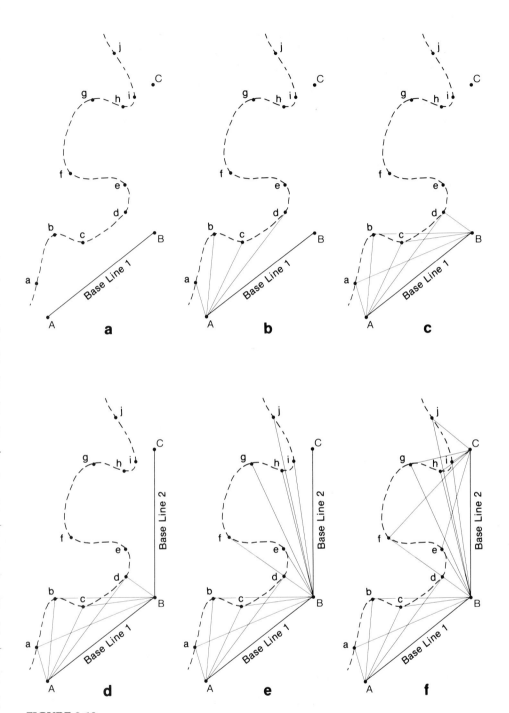

FIGURE 2.18
Step-by-step procedure for plane-table mapping using the triangulation method. (a) The dashed line represents a stream and the triangulated points **a** through **j** mark the stream's position. (b) Begin by establishing a base line (**Base line 1**) and sight to points along the stream (**a,b,c,d**). (c) Then, move to the other end of the base line and sight-back to the same points (**a,b,c,d**). (d) Establish a second base line (**Base line 2**). (e) Sight to other critical points along the stream (**e,f,g,h,i,j**). (f) Move to the other end of **Base line 2** and sight to the previously sighted points (**e,f,g,h,i,j**). Continue this process until the open traverse has been completed.

10. Note: Depending on the size and shape of the traverse, it may be necessary to establish a second base line (figure 2.18d) to map all of the triangulation points. This can be done while positioned at point B. Use point B as the beginning point for the second base line, and follow the same procedure used to create the first base line. Once a second base line has been determined, additional triangulation points can be located in the same manner as outlined in steps 6, 7, and 8 (figures 2.18e and 2.18f).

3

Topographic Maps

INTRODUCTION

Topographic maps are part of a widely used series of maps designed and produced under the direction of the United States Geological Survey (USGS). Topographic maps have been produced by the USGS in one form or another since 1879. While evolution in design has occurred over the years, the overall intent of bringing together physical and cultural features on an accurately surveyed base map has remained the central purpose of the topographic map.

Each map in a USGS series conforms to established specifications for size, scale, content, and symbolization. Except for maps that are formatted on a county or state basis, USGS quadrangle-series maps cover areas bounded by parallels of latitude and meridians of longitude.

MAP SCALE

Topographic maps are classified generally by publication scale, and each scale fulfills a range of map needs. Map scale defines the relationship between the dimensions of the features as shown on the map and the dimensions as they exist on the Earth's surface. Scale is generally stated as a ratio or fraction—1:24,000 or 1/24,000. The left side of the ratio, or the numerator, is customarily 1, and represents map distance. The right side of the ratio, or the denominator, a larger number, represents horizontal ground distance. Thus, the scale 1:24,000 indicates that any unit, such as 1 inch or 1 cm on the map, represents 24,000 of the same unit on the ground.

Large-scale maps, such as 1:24,000, portray relatively small areas of the Earth's surface in considerable detail. They are especially useful for highly developed areas, or for rural areas where detailed information is needed for engineering planning, siting of mining operations, drilling, or similar purposes.

Intermediate-scale maps, ranging from 1:50,000 to 1:100,000, cover larger areas and are especially suited for land management and planning.

Small-scale maps, such as 1:250,000, 1:500,000, and 1:1,000,000, cover very large areas on a single sheet and are useful for comprehensive views of extensive projects or for regional planning.

QUADRANGLE SYSTEM OF MAP LAYOUT

Although publication scales have changed, the system of subdividing areas for mapping purposes is the same as originally devised in 1882. The universal coordinate lines of latitude and longitude are used to form the boundaries of four-sided figures called quadrangles (figure 3.2, tinted areas). Quandrangles are the units of area adopted for topographic mapping. Each map sheet is the map of a quadrangle, and the maps themselves are called "quadrangle maps."

The system of subdivision provides quadrangles of different sizes suited to mapping areas at various scales (figure 3.1). Generally, the larger quadrangles are bounded by degree lines of latitude and longitude; smaller quadrangles are derived by subdividing the larger ones (figure 3.1). Thus, a 1-degree quadrangle comprises four 30-minute quadrangles; a 30-minute quadrangle, four 15-minute quadrangles; and a 15-minute quadrangle, four 7.5-minute quadrangles. Because the meridians (vertical lines of longitude) converge toward the north in the Northern Hemisphere (figure 3.2), the shape of a quadrangle is actually that of a trapezoid, although the variation from a true rectangle is very small near the equator, but increases as you progress poleward.

Not all quadrangles, however, have the same dimensions in latitude as in longitude. For example, the north-south dimension (measuring between lines of latitude) of Alaska maps in the 1:63,360-scale series is 15 minutes, but the east-west dimension (measuring between lines of longitude) varies from 20 to 36 minutes, depending on the latitude. The larger dimensions in longitude are needed to avoid excessively narrow maps in the higher latitudes (see upper tinted area in figure 3.2). The regular quadrangle shape is also modified in many areas to include islands and small coastal features. The 1:100,000-scale series is produced in a 30-minute by 1-degree format, and the 1:250,000-scale series in a 1-degree by 2-degree format.

Although the quadrangle size in terms of latitude and longitude is generally constant, the ground area covered varies considerably. Quad-

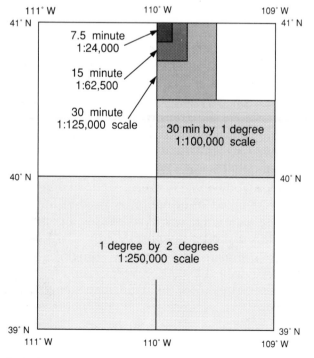

FIGURE 3.1
Comparative coverage of quadrangles from U.S. Topographic Map Series.

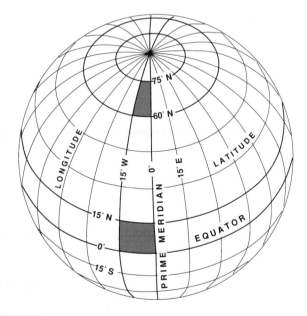

FIGURE 3.2
The geographic coordinate system is composed of lines of latitude drawn parallel to the equator and lines of longitude extending from pole to pole. Notice that lines of longitude converge at the poles. Therefore, a map of 15° by 15° at the equator covers a greater area than a map of 15° by 15° covers at the Arctic Circle (compare gray areas).

rangles in the northern part of the Northern Hemisphere cover a smaller area than those in the southern part—again, see figure 3.2. Therefore, in the figure, the 15-degree by 15-degree area shown near the Arctic Circle does not cover as much surface area as the 15-degree by 15-degree area near the equator. This is the direct result of converging lines of longitude.

There is a systematic relationship between the size of a quadrangle, and the scale at which a map of the quadrangle is printed. If the publication scale is doubled and the quadrangle size is halved, the paper area required for printing the map remains the same. Thus, the regular ratios between quadrangle sizes and publication scales (figure 3.1) make it possible to print maps in the first three sizes on paper of the same dimensions, which is a convenience in production, shipping, using, and filing. It is also useful to be able to measure distances on different maps and convert to other scales by a simple whole-number multiplication or division.

The regular sequence of scales was not followed, however, for the 7.5-minute series. Instead of twice the 15-minute scale of 1:62,500 (which would have been 1:31,250), two scales were adopted—initially 1:31,680 and later 1:24,000. The 1:24,000 scale remained standard for this series of maps until the shift to the metric system and the 1:25,000 scale. A larger sheet of paper is needed for the 1:24,000 scale, but it is a useful scale when the maps are employed directly as planning bases. Also, the even ratio of inches to feet (1 inch = 2,000 feet at the scale of 1:24,000) is convenient in engineering work based on our customary system of measurement. With the changeover to the metric system, a scale of 1:25,000 (1 cm = 0.25 km) is appropriate.

MAP GRIDS

USGS topographic maps feature several grid systems. In addition to the standard longitude and latitude grids, most topographic maps contain a State Plane Coordinate System (SPCS), a Universal Transverse Mercator (UTM) grid, and, in some areas, the United States Public Lands Survey (USPLS) system. All of these are included on the maps to broaden their applicability for many users.

United States Public Lands Survey

One of the most visible grids seen on many topographic maps is that of the Public Lands Survey. The system is indicated by red lines that break up the surface of the map into numbered rectangles.

The Public Lands Survey applies to lands west of the Appalachians, or lands that were not part of the original 13 colonies. The survey was developed in the late eighteenth century to assure proper titling of land claimed by settlers. Some areas were not surveyed, or were incompletely done, in regions that were previously held by foreign governments before becoming part of the United States or its territory. The system was not implemented in some areas, in order to protect private land claims. Also, it was not done in many inaccessible locations such as rugged mountains or swamps.

The outcome was a systematic survey which divided the land into easily defined rectangular blocks. Township-range blocks are identified using an alphanumeric system. Townships are referenced in a north-south direction; ranges are identified in an east-west orientation.

Township and range block units are initiated by the intersection of a meridian, called the *principal meridian*, and a parallel called the *base line*. To compensate for curvature of the Earth, a number of intersection points are used to initiate different block units throughout the United States. The system of parallels and meridians is illustrated in figure 3.3.

Figure 3.4 illustrates the land-division scheme. The largest unit is the *township-range block*. These are laid out as approximately 6-mile by 6-mile blocks (one is marked T 2 S, for example). Each block is divided into 36 smaller rectangles called *sections*.

Townships are identified by an alphanumeric sequence. In the example, T 2 S designates the second township south of the established base line of the survey. It is read "Township-Two-South." Ranges use a similar sequence. The range designated R 3 E is the third range east of the established principal me-

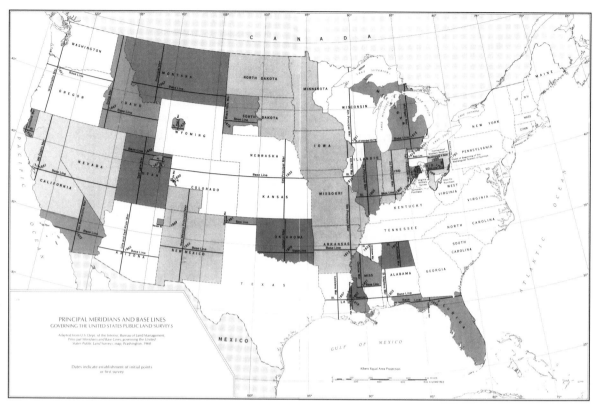

FIGURE 3.3
U.S. Public Lands Survey (USPLS) system. Origin points for the township and range system are keyed to a sequence of land acquisition which eventually completed the formation of the conterminous United States. (USGS)

ridian of the survey. It is read "Range-Three-East."

Sections are identified in the margins of the map in a combination of letters and numbers printed in red. Sections are numbered between 1 and 36, beginning with the most northeastward section (upper right corner) and proceeding to the west in sequence to the end of the row. The numbering continues on the next row, this time toward the east. This sequence continues in a zigzag pattern until arriving at block 36 in the southeast corner.

Each section can be further subdivided into half sections, quarter sections, half-quarter sections, quarter-quarter sections, and lots (figure 3.4). Aerial photographs and satellite imagery frequently capture land-use patterns that mirror this land-division scheme.

The location description for a parcel of land is relatively simple. The southeast land parcel observed in Section 29 (figure 3.4) is described as follows: E1/2, SE1/4, Sec. 29, T 2 S,

R 3 E. Put in a different way, the address of a location is hierarchical. It is initiated by defining the part of the section, then the section, and finally the township and range. If required, you could also cite the name of the principal meridian used as the origin of the system—for example, Wilamette.

Universal Transverse Mercator (UTM) Grid

A less-conspicuous but equally important grid system used on topographic maps for establishing location is known as Universal Transverse Mercator, or UTM. On the margins of topographic maps you can observe small, light-blue tick marks accompanied by identifying numbers. The numbers represent the horizontal or vertical distance, in meters, that the grid line lies from a false origin located to the south or west of its plotted location. When printed, the numbers are mostly abbreviated; the last three

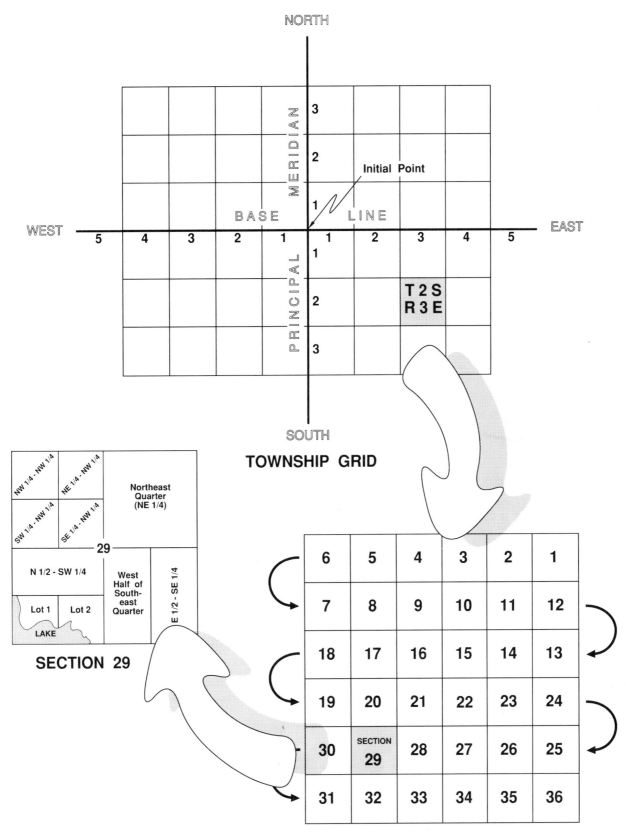

FIGURE 3.4

Land subdivision in the U.S. Public Lands Survey (USPLS). One township and range defines a 36-square-mile block of land. The block is divided into 36 sections that are each 1 square mile in area and are numbered 1–36. Each section can be further subdivided into quarter- or half-sections. These are identified using cardinal directions as references—N/S/E/W or NE/NW/SE/SW. (After Thompson 1979)

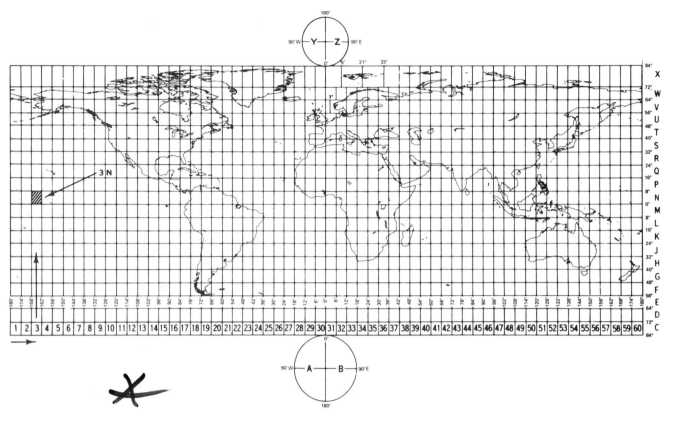

FIGURE 3.5

Universal Transverse Mercator (UTM) grid zones designation. Illustrated on an expanded equidistant cylindrical projection are the sixty 6° zones of the UTM system. Each zone is numbered and is further subdivided into latitudinal parcels identified with a letter. Circular parcels at the bottom and top of the map represent the Universal Polar Stereographic (UPS) zones used to map the polar regions. (USGS)

digits (000) are missing. Consequently, the distances actually are greater than you might have thought upon cursory evaluation of the map. A section of the margin on a topographic map is illustrated in figure 3.7.

The UTM grid is created through the use of a cylindrical projection system (figure 3.5). In this case, the cylinder is laid parallel to the plane of the equator. The purpose of the orientation is to create a series of zones, each centered on a meridian. The walls of the cylinder will contact the surface of the sphere at two locations, each of which is 3° away from the central meridian. This method of plotting geographic data is reasonably accurate in terms of shape, area, direction, and distance. The Transverse Mercator grid is characterized by having two standard meridians.

The outcome of transforming data from the sphere of the Earth to the plane of the map is a cylindrical map having 60 parallel zones, each

covering 6° of longitude and 164° of latitude. The north and south borders of the grid lie at lat. 84° N and lat. 80° S (figure 3.5).

After the zones have been plotted, each is assigned a number from 1 to 60. The zone numbers increase eastward, beginning at long. 180° (left to right in figure 3.5). The zones are further divided into two parts—north and south—each pertaining to a hemisphere of the Earth. To represent the conterminous United States in its full west-to-east extent, 10 zones are required, numbered 10–19.

For complete coverage of the world, it is necessary to add two polar projections to the UTM system. They are also shown in figure 3.5, above and below the 60-zone map. The Northern Hemisphere polar projection extends poleward from 84° N and is divided into Western and Eastern Hemisphere zones (Y and Z, respectively). In the Southern Hemisphere, the polar projection extends from 80° S to the South

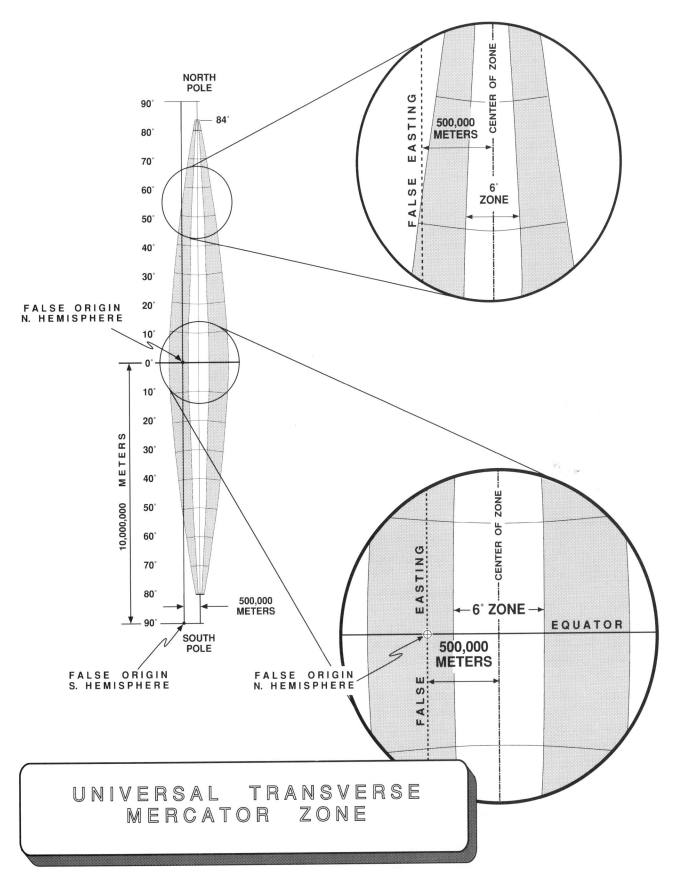

FIGURE 3.6

Universal Transverse Mercator zone. A single zone extends from lat. 84° N to lat. 80° S. Each zone is 6° wide. False origins for the zone lie at two different locations. For the Northern Hemisphere, the false origin is 500,000 meters west of the central meridian (center of zone), on the equator. The Southern Hemisphere-portion has its false origin at 10,000,000 meters south of the equator, and 500,000 meters west of the central meridian. Calculations of eastings and northings are made from one of these origins.

Pole, and is also divided into hemisphere zones (A and B, respectively).

In effect, the UTM grid is a Cartesian system. Each line in the grid marks distance horizontally and vertically from an origin point (figure 3.6). The origin point is termed *false*. A false origin is one which lies outside of the zone being defined by the grid. The false origin for the Northern Hemisphere-portion of a zone lies 500,000 meters to the west of the central meridian of the zone, and along the equator. The false origin of the Southern Hemisphere-portion of a zone is located 10,000,000 meters to the south of the equator, and 500,000 meters west of the central meridian of the zone. Therefore, each transverse zone has two origin points for the measurement and definition of the grid (figure 3.6).

Significantly, deriving coordinates for locations within a UTM zone is a reasonably simple task. A location is obtained by making an *easting* measurement (the distance, in meters, east of the false origin), followed by a *northing* measurement (the distance, in meters, north of the false origin). Another way of expressing this is "read right, then up."

Figure 3.7 contains a section of a topographic map. We have added a grid using the UTM marks along the map margins. (If you need to draw UTM grid lines on your own topographic maps, it is best to use a nonphotographable light-blue pencil.)

On this map, the town of Rock Springs, Maryland is near the center. To establish the UTM coordinates for Rock Springs, first identify the closest 1,000-meter northing and easting that lie south and west of the town, respectively. In this example, they are 4,396,000 m N and 400,000 m E. Next, subdivide the 1,000-meter intervals straddling the location (Rock Springs) into 10 equal parts of 100 meters each (this is the small 10 × 10 grid we have superimposed over Rock Springs). Determine how many meters north and east the town center is from the previously identified 1,000-meter grid lines. In our example, the northing and easting would be 4,396,500 m N and 400,700 m E, respectively.

For further accuracy, divide the closest 100-meter northing and easting into 10 equal parts. Read the interval of the number closest to the town center. On 1:24,000-scale maps, this is as close as you can accurately plot. In the cited example, the final northing and easting are 4,396,540 m N and 400,720 m E.

A mathematical way to determine northings and eastings is to locate the 1,000-meter northing closest to Rock Springs, as was done in the first example. Using an engineer's scale, measure the distance from the 1,000-meter grid line to the town center. Divide this distance by the measured distance between the two 1,000-meter grids bounding the town-center location. Multiply the result by 1,000 meters, and add it to the first 1,000-meter northing.

In our example, the mathematics for the northing are:

N = grid line closest to Rock Springs that does not exceed the distance to the location: 4,396,000 m N

md = measured distance from Rock Springs to the 1,000-meter grid: 0.675 inches

mg = measured distance between the two 1,000-meter northings that bound the point of location: 1.25 inches

Now, plug these values into the formula:

$$\text{northing} = N + \frac{md}{mg} \times 1,000 \text{ m}$$

$$\text{northing} = 4,396,000 \text{ m N} + \frac{0.675 \text{ in}}{1.25 \text{ in}} \times 1,000 \text{ m}$$

$$\text{northing} = 4,396,000 \text{ m N} + 0.54 \times 1,000 \text{ m}$$

$$\text{northing} = 4,396,000 \text{ m N} + 540 \text{ m}$$

$$\text{northing} = 4,396,540 \text{ m N}$$

To establish the easting, repeat the procedure, using easting readings.

To complete the UTM description of Rock Springs coordinates, combine the northing and easting readings with the zone number and

FIGURE 3.7
UTM grid on a USGS quadrangle map. On U.S. topographic maps, blue tick marks accompanied by dual-type-size identifying numbers (example: [4]01 at top center map border) indicate the UTM system. In this illustration, we have added UTM grid lines to a section of a quadrangle, using the tick marks as references. (USGS)

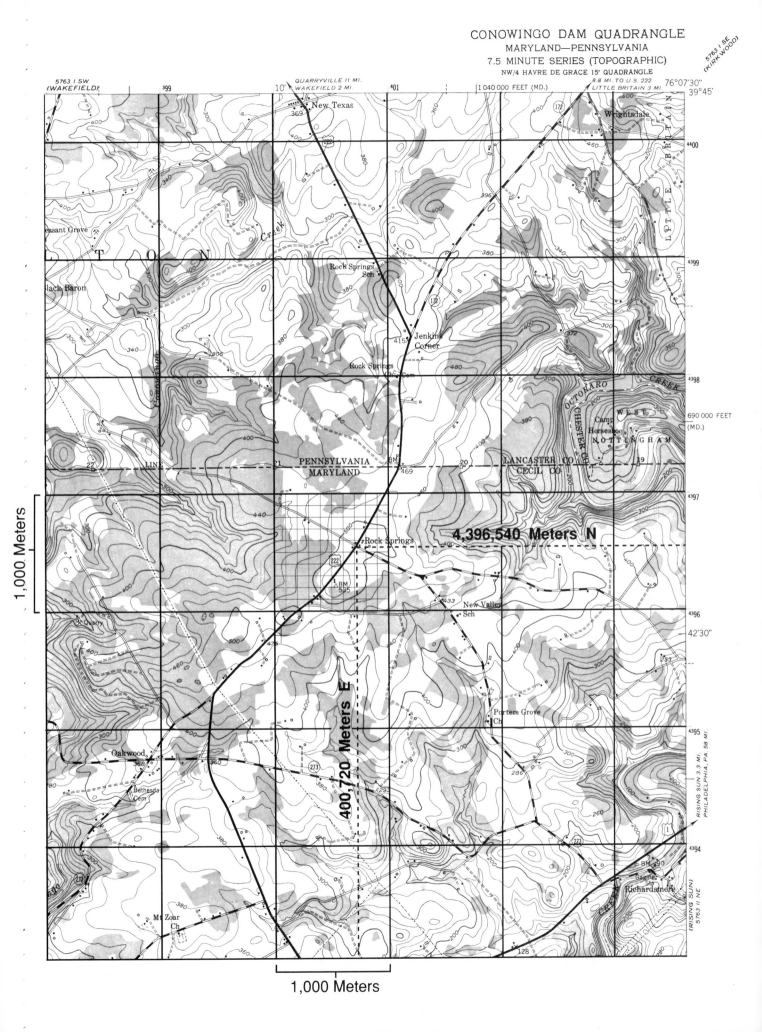

1,000 Meters

4,396,540 Meters N

400,720 Meters E

1,000 Meters

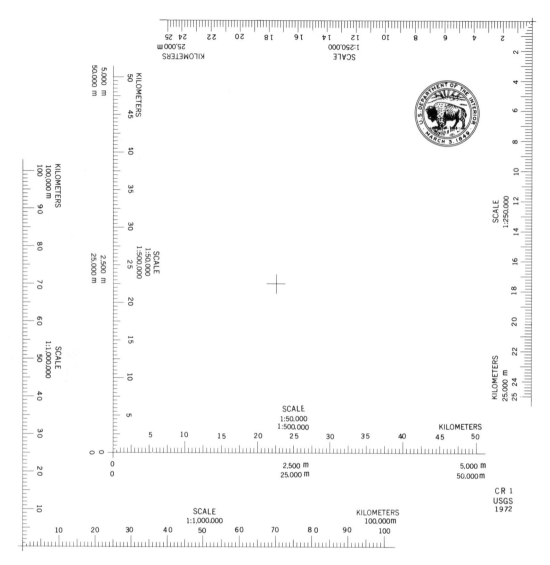

FIGURE 3.8a
Metric coordinate reader. Frequently referred to as a "roamer," the metric coordinate reader is an instrument designed to facilitate measurements on topographic maps. Roamers are usually produced on transparent plastic so you can conveniently see reference lines/points on the map base. Increments on the roamer are designed to measure distances at a variety of scales. Students can make their own roamers by photocopying figures 3.8a and b at 100% size onto transparencies. Illustrated is the CR-1 template, which is designed for general use with metric scales. (USGS)

zone half. The final coordinate description is: 400,720 m E, 4,396,540 m N, Zone 18, N.

The location of coordinates is expedited by using an instrument called a *roamer* (figures 3.8a and 3.8b). Roamers are clear plastic templates having a surface embossed with conveniently marked intervals for measuring distances in meters. They can be easily moved about the surface of a map for quick linear measurement. Depending on the scale of the map, the grid marks may be subdivided into units as small as 10-meter intervals.

State Plane Coordinate Grid

Topographic maps are characterized by a third grid system, known as the *State Plane Coordinate System (SPCS)*. Each state in the United States has its own State Plane Coordinate System (figure 3.9). Note how most states are divided into zones. States having a greater east-west dimension than north-south are divided into zones extending in an east-west direction (Kansas, for example). States that are elongated in the north-south dimension are broken into north-

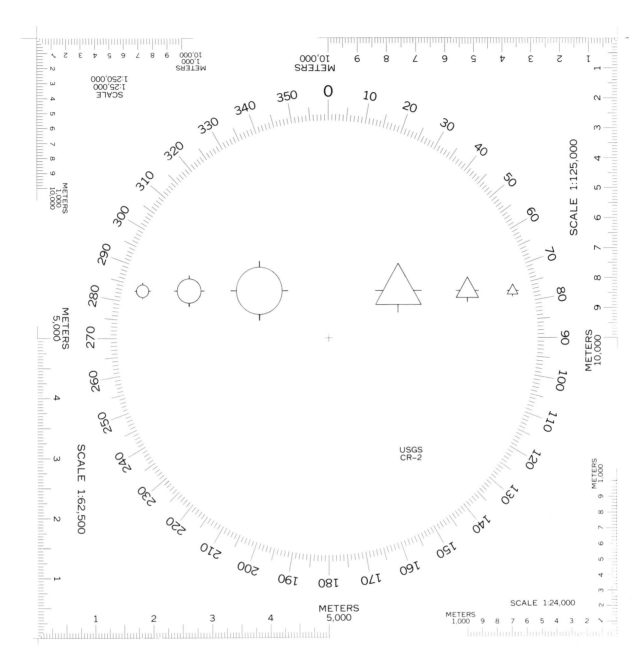

FIGURE 3.8b
Metric coordinate reader. Similar to the template in figure 3.8a, this template (CR-2) is designed for scales used most frequently by the USGS. (USGS)

south-oriented zones (Arizona). Zone boundaries follow the outlines of counties, to avoid dividing up counties.

The creation of these zones is based upon the use of different projection systems. A conic system (Lambert conformal) is employed to create the grid for states elongated east-west, such as Montana (figure 3.10). A cylindrical system (Transverse Mercator) is used to develop zones for states extending north-south, such as Wyo-

ming. In each instance, the projection was selected to minimize plotting errors resulting from the curvature of the Earth's sphere. Subdivision of a state into different zones is for the same purpose—minimizing plotting errors. Therefore, surveyors can use the SPCS to obtain the coordinates for places being surveyed.

The State Plane Coordinate System creates a rectangular grid. Marks that enable you to draw the grid exist as black tick marks in the

FIGURE 3.9
State Plane Coordinate System (SPCS). Layout of system is illustrated for the entire United States. Some states are divided into zones extending north-south through the state; others are divided into east-west-extending zones.

margins of topographic maps. Each tick mark is identified by a number, such as "690,000 feet." Like UTM, this system is a Cartesian type; consequently, the numbers increase from left to right across the map and from the bottom toward the top. In this system, the numbers represent feet, not meters, east and north from a false origin.

Figure 3.11 shows a section of a topographic map. Using the marginal tick marks, we have added a series of lines defining the SPCS. Let us read the coordinate location for Rock Springs. The location is calculated in the same manner as described for the UTM grid system, except northings and eastings are given in feet instead of meters. Therefore, the Maryland coordinates for Rock Springs are "1,036,505 feet E, 686,586 feet N; Maryland" (state name), and followed by the zone identi-

fier if more than one zone exists for the state (but Maryland only has one zone). Many states do contain two or more zones; California has six. You should indicate into which zone the place being referenced falls: West, East, North, Central, South, etc.

SERIES MAPS AND SPECIAL MAPS

A *map series* is a set of maps that conforms generally to the same specifications, and covers an area or a country in a systematic pattern. The maps of a series have the same format, quadrangle size, and system of symbolization, and usually the same scale. Adjacent maps of a series can be combined to form a single large map; the features will match across the joined edges because the symbols and treatment are the same.

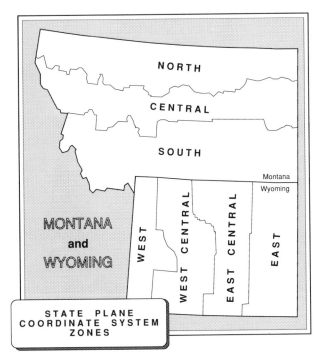

FIGURE 3.10
State Plane Coordinate System in Montana and Wyoming. A conic projection was used to create the division of Montana, whereas Wyoming was divided using a cylindrical system.

Map series may be designated by special titles, quadrangle sizes, or publication scales (figure 3.12). A series may be referred to by the quadrangle size only if the quadrangles have the same dimensions in both latitude and longitude; for example, the "15-minute series" means that all the quadrangles in that series measure 15 minutes by 15 minutes. Otherwise, the series must be identified by its scale, such as the Alaska 1:63,360-scale series.

Although the great majority of the U.S. Geological Survey's published maps are series maps, some are published at special scales, cover special areas, or have some unique feature so that they cannot be included in a series. These are called "special maps" and are listed individually in map indexes. National mapping requirements are the basis for the National Mapping Program, which includes a number of individual map series.

The following are representative of *topographic* series prepared, revised, and distributed by the Geological Survey:

7.5-minute	United States
Puerto Rico, 7.5-minute	International Map
15-minute	of the World (IMW)
Alaska 1:63,360-scale	National Park
1:250,000-scale	Antarctica
State	Miscellany
	(special areas)

Maps in the following series are generally not revised or reprinted. Copies of these maps, and maps replaced by revisions, are held in the historical file, and black-and-white copies are supplied on request at the cost of reproduction:

30-minute	Alaska reconnaissance
1-degree	Metropolitan Area

Complete coverage of the United States exists, or is planned, for the 1:25,000-scale, State, and International Map of the World (IMW) series. Complete coverage at 1:24,000 scale is planned for the conterminous States and Hawaii. Complete 1:63,360-scale coverage is planned for Alaska (1:25,000-scale coverage in Alaska is provided in areas of special need). The amount of coverage at 1:62,500-scale depends on demonstrated need and demand; however, if initial publication is at 1:62,500, manuscripts are prepared at 1:24,000 with accuracy, content, and contour interval suitable for future publication in the 7.5-minute series. Copies of the manuscripts are held as open-file material, and made available as black-and-white copies.

SELECTION OF MAPPABLE FEATURES

The biggest problem in map presentation is to make the best use of available space. If a map is crowded with too many lines and symbols, it will be unreadable; yet the amount of information that might be useful or desirable is almost unlimited. The cartographer must select the features that will be most valuable to the map user. The smaller the map scale, the more critical and difficult the problem of selection becomes.

Quadrangle maps produced by the USGS are designed to be used for many purposes. Scales, contour intervals, accuracy specifications, and features that are shown on the maps

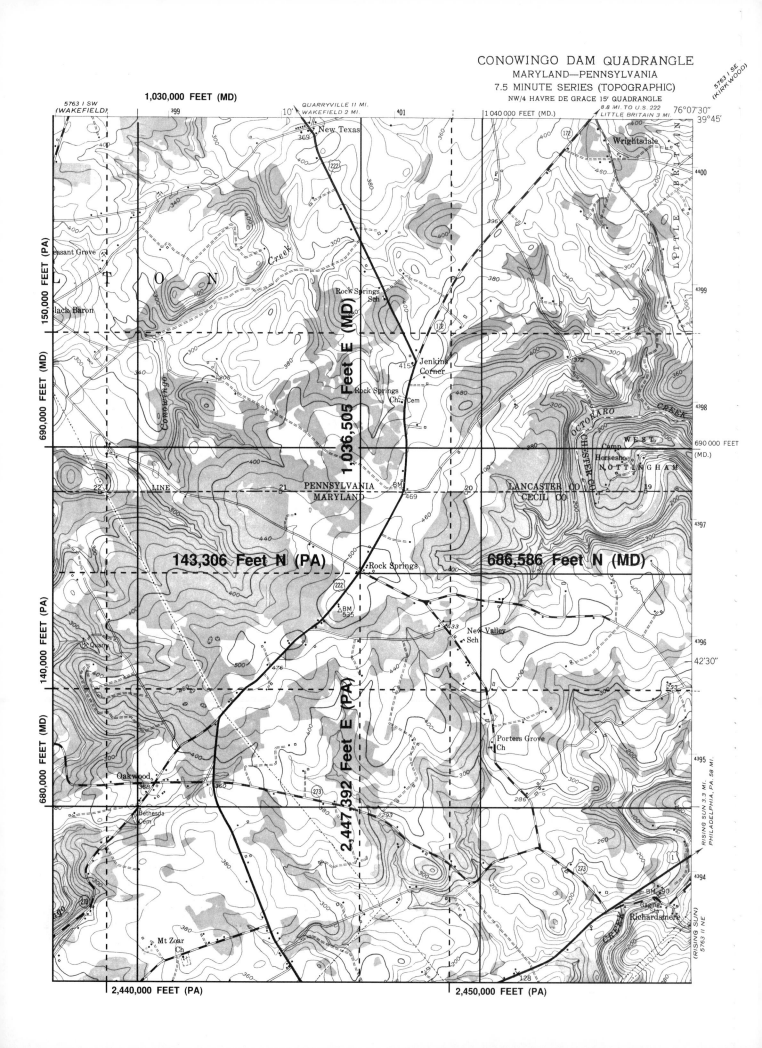

FIGURE 3.11
SPC grid on topographic quadrangle. Locations are identified by giving an easting and northing in feet. We have filled in the SPC grid to facilitate the location of Rock Springs, Maryland. Since Rock Springs lies near the Maryland/Pennsylvania border, the SPC system for either state could be used to locate the town. The coordinates for Maryland are 1,036,505 ft E, 686,586 ft N; Maryland. If the Pennsylvania SPC is used, the position is 2,447,392 ft E, 143,306 ft N; Pennsylvania, S (for south zone). (USGS)

have been developed gradually to satisfy the requirements of government agencies, industry, and the general public. Because these maps serve a wide variety of uses, they are called general-purpose maps.

The functions that a map is intended to serve determine which features should be mapped. However, other factors must be considered before it is decided what specific features actually can be shown. Among the most important considerations are the permanence of the features, the cost of compiling the information, and map legibility.

The legibility requirement means that small features must be represented by symbols that are larger than true scale size. For example, a paved secondary road, which normally ranges 20 to 40 feet wide on the ground, is shown as a double line symbol with a width of 0.025″ on a 1:62,500-scale map. If the 0.025″ wide line symbol were scaled up to actual ground size, it would be 130 feet wide! Or, conversely, if the line symbol were drawn to scale for a 30-foot-wide road to be shown on a 1:62,500-scale map, it would only be about 0.006″ wide. Buildings and other structures also are depicted by symbols that may be larger than the scale size of the features. If smaller objects were represented at their true scale size, the symbols would be too small to be legible.

In mapping congested areas, some features are omitted because symbols for these features generally occupy more space on the map than their actual size merits.

The extent to which some kinds of map features are shown is determined partly by the cost of compiling the information. Aerial photographs are the source of most map information, but features that cannot be identified on photographs must be mapped by field methods, an expensive procedure. As an example, not all public-land section corners are shown; they are too small to be seen on aerial photographs, and the cost of mapping them using field surveys would be excessive.

Series	Scale	1 Inch Represents Approximately	Size (latitude x longitude)
Puerto Rico 7.5-minute	1:20,000	1,667 feet	7.5 x 7.5 minutes
7.5-minute	1:24,000	2,000 feet (exact)	7.5 x 7.5 minutes
7.5-minute	1:25,000	2,083 feet	7.5 x 7.5 minutes
7.5 x 15-minute	1:25,000	2,083 feet	7.5 x 15 minutes
USGS/DMA 15-minute	1:50,000	4,166 feet	15 x 15 minutes
15-minute	1:62,500	1 mile	15 x 15 minutes
Alaska 1:63,360	1:63,360	1 mile (exact)	15 x 20 to 36 minutes
County 1:50,000	1:50,000	4,166 feet	County area
County 1:100,000	1:100,000	1.6 miles	County area
30 x 60-minute	1:100,000	1.6 miles	30 x 60 minutes
U.S. 1:250,000	1:250,000	4 miles	1 x 2 or 3 degrees
State maps	1:500,000	8 miles	State area
U.S. 1:1,000,000	1:1,000,000	16 miles	4 x 6 degrees
U.S. Sectional	1:2,000,000	32 miles	State groups
Antarctica 1:250,000	1:250,000	4 miles	1 x 3 to 15 degrees
Antarctica 1:500,000	1:500,000	8 miles	2 x 7.5 degrees

FIGURE 3.12
Topographic map series comparisons. This table lists the topographic map series, associated scale, and longitude and latitude extent. (USGS)

Not only the original compilation cost, but the cost of keeping the map up-to-date as well, is considered in deciding which features to map. Generally, the more features that are depicted, the more quickly the map becomes out-of-date. Cultural features are especially subject to change. If the maps are to have a reasonably long useful life, the features portrayed must be restricted, to some extent, to relatively permanent objects.

Many kinds of features are shown on some maps, but omitted from others, because of their landmark character. In this sense a landmark is an object of sufficient interest in relation to its surroundings to make it stand out. For example, buildings may be considered landmarks when they are used as schools or churches, or when they have some public function. They may be landmarks also because of their outstanding size, height, or design; or they may be landmarks because of their history, such as old forts or the birthplaces of famous people.

The same principle is applied to features other than buildings. The adjacent area always is considered in relation to the object in deciding whether it qualifies as a landmark. Where map features are few, objects that would not be shown in more congested districts may be mapped as landmarks.

MAP SYMBOLS, COLORS, AND LABELS

A topographic map, as distinguished from other kinds, portrays by some means the configurations and elevations of the terrain—the shapes into which the Earth's surface is sculptured by natural and sometimes manmade forces. USGS topographic maps usually represent elevations and landforms by means of contour lines. Other features are shown by a variety of conventional signs, symbols, lines, and patterns printed in appropriate colors and identified by names, labels, and numbers. The topographic map symbol chart (in the back of the book) shows the standard symbols used on the topographic maps of the USGS.

The features shown on quadrangle maps are divided into three general classes, each printed in a different color. Relief features—topographic or hypsometric information—are printed in brown. Water features are shown in blue, and cultural features—manmade objects—in black. The system of division is not rigid, however. Levees, earth dams, and some other manmade features are also topographic features, but they are printed in brown, not black.

Besides the colors used for the three main classes of features, green is used to show woodland—timber, brush, vineyards, and orchards—and red is used to show public-land subdivision, built-up areas, and the classification of the more important roads.

Linear features are represented by lines of various weights (widths) and styles (solid, dashed, dotted, or a combination). Structures or individual features are portrayed by a system of pictographs or symbols. The symbols originated as plan views (bird's-eye views) of the objects they represent, and they retain something of this character although they are now formalized. The building symbol, for example, is a solid or open square. The railroad symbol is a line with evenly spaced cross-ties. The dam and levee symbols look approximately like dams or levees as seen from the air.

Because lines and symbols cannot represent map information completely, they are supplemented by the names of places and objects. Notes are added to explain some features that cannot be depicted clearly by symbols alone. In mapping topographic features the information portrayed by contour lines is supplemented by elevation figures. Letters and numbers are essential to map interpretation, but because they tend to obscure other map information, they are selected and positioned carefully to minimize interference with other detail.

RELIEF INFORMATION

The two main reasons for showing relief information on maps are (1) to furnish coordinated data for computing problems involving terrain dimensions, and (2) to present a graphic picture of the ground surface. The two objectives are related by distance, and sometimes they may conflict.

For engineers or scientists who are interested in exact measurement, topographic maps furnish dimensional information about elevations, areas, grades, and volumes. The approximate elevation of any point can be read directly or interpolated from contours. A series of elevations on a line determines the grade or profile of the line, and areas and volumes can be computed by combining line profiles in various ways. The relief information shown by contours is sufficient for calculating the storage capacity of a reservoir, the area of a stream or river drainage basin, or the volume of earth to be moved in a large road cut or fill.

On the other hand, many persons who use maps are not concerned with the exact ground elevations, but are more interested in the general appearance and shape of the land. For them, contours are the graphic means of visualizing the terrain and an aid in locating positions on the map.

Contour lines are the principal means used to show the shape and elevation of the land surface. Other means are spot elevations, relief shading, hachures, and pattern symbols for special kinds of relief features that are not suited to contouring.

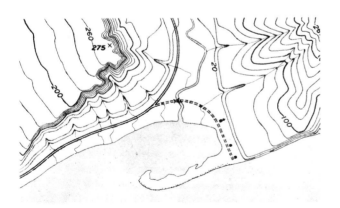

FIGURE 3.13
Relationship between physical terrain and a contour map. Contours on the map (bottom) represent the ground configuration in the perspective rendering (top). (USGS)

Contours are lines of equal elevation. They always are continuous lines, and may form closed loops. A contour is an imaginary line on the ground, every point of which is at the same elevation above a specified datum surface (mean sea level, for topographic maps of the USGS). The *contour interval* is the difference in elevation between adjacent contours. The contour interval, together with the spacing of the contour lines on the map, indicates the slope of the ground. On steep slopes the lines are spaced more closely than on gentle slopes.

The basic characteristics of contours are illustrated in figure 3.13, which shows an oblique view of a river valley and the adjoining hills, and the same features as shown on a topographic map. Note that when contours cross a stream they form a "V" pointing upstream. Furthermore, closed contours can aid in quick identification of hill tops and ridges. Valley bottoms are easily located by more widely spaced contours and the frequent presence of a stream. Closely spaced contours indicate steep terrain. When looked for, these few characteristics allow quick interpretation of map relief.

Contour Interpolation

There are times when you must construct contour coverage for areas that have not been previously mapped or for which you do not have appropriate contour data. Topographic-map coverage may be unavailable at the desired scale and contour interval, or map coverage may be completely absent. Whatever the reason, you may have to modify an existing map (sketch-in additional contours), or you may have to create a new contour map, to meet mapping objectives. Both are achieved through the process of interpolation.

While there are various interpolation schemes, the most widely used for creating elevation contours is *interpolation based upon a linear gradient*. This technique assumes that the slope is constant between two adjacent elevation points (or between two contour lines, if an existing contour map is being modified). Therefore, a line drawn between the two elevation points can be divided into segments of equal length. The number of segments depends upon the contour interval and the difference in elevation between the two points.

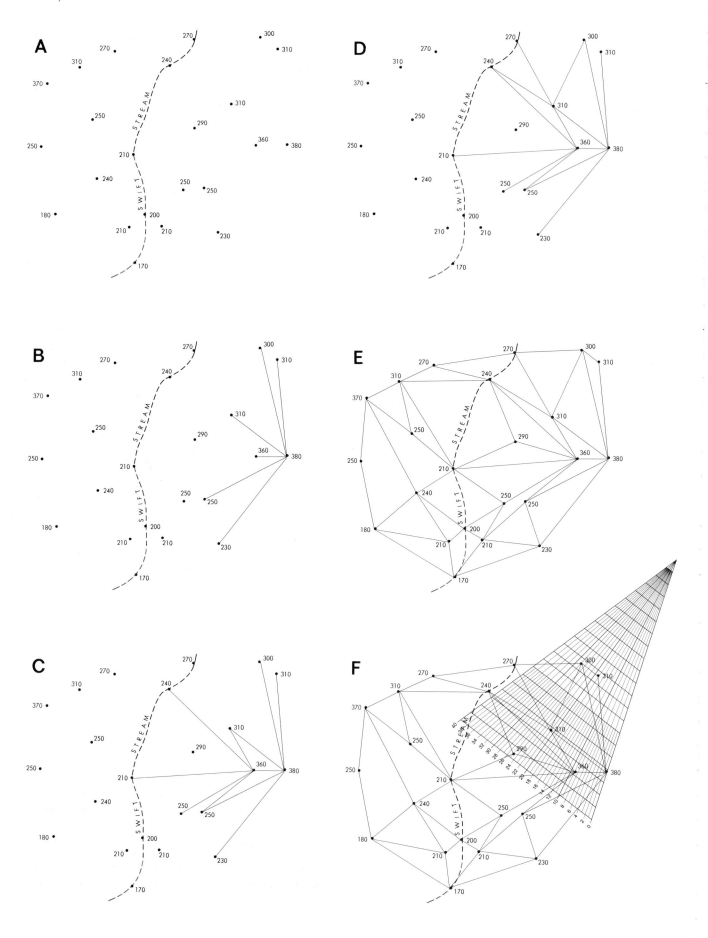

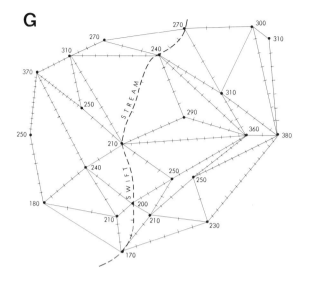

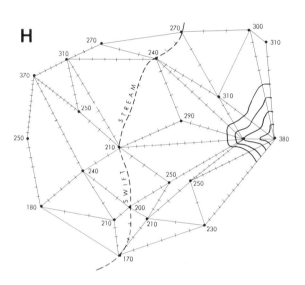

For example, figure 3.14A locates spot elevations obtained by field surveying. To establish the location of contour lines, start on the right side of Swift Stream, and connect all of the elevation points with straight lines to form a series of triangles (figure 3.14B). (Use a non-photo-blue pencil such as a Berol 919.) Start with the highest elevation (380′) and extend lines to lower elevation points. Then, proceed to the second-highest elevation point on the right side of Swift Stream (360′ in figure 3.14C) and extend lines to lower elevation points. (Note: do not cross previously drawn lines, and do not extend lines across streams or water bodies.)

Move to the third-highest point (310′ in figure 3.14D) and repeat the process. Continue to successively lower points until all points are connected to form triangles, with an elevation point at each apex. Close any incomplete triangles. Move to the other side of the stream and repeat the procedure (figure 3.14E).

When all triangles are formed, use the *linear divider grid* (figure 3.14F) to divide each triangle leg into an appropriate number of segments, based upon the contour interval. For this example we have selected a contour interval of 10 feet. In figure 3.14F the elevation difference between the points at 380 ft and 310 ft is 70 ft. Therefore, using a 10-foot contour interval, the line from 380′ to 310′ (the interior 310′ point, northwest of 380′) is divided into 7 equal segments (70 ft/10 ft = 7 segments).

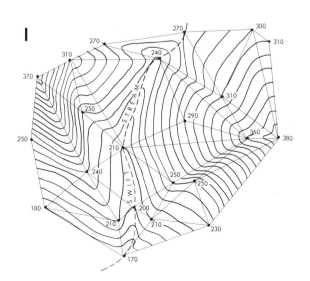

FIGURE 3.14

Contour interpolation. This illustration presents a step-by-step procedure for contour interpolation from spot elevations (A). (B) Begin with the highest point on the map. Extend lines outward from the highest point to other nearby points. (C,D) Repeat the process for each progressively lower point until the entire map is covered with a series of triangles (E). (Note: Do not cross previously drawn lines, and do not extend lines across streams or bodies of water.) (F) Use a linear divider grid to segment each triangle side in accordance with the contour interval and the elevation difference between the end points of the triangle. (G) Repeat this procedure until all triangle sides have been divided. (H) Beginning with the highest point, draw contour lines by connecting tick marks that have the same elevation. (I) Continue until all contour lines have been constructed.

On a light table, place the map over the linear divider grid (figure 3.14F). Keep the line that you drew between 380′ and the 310′ point to the northwest parallel to the parallel lines on the linear divider grid. Slide the map over the grid until the 380′ and 310′ points are positioned with 7 segments between them (or a multiple of 7; we used 3 × 7, or 21 segments). Now divide the line into 7 equal segments (every third grid line) with tick marks (figure 3.14G) (use the nonphoto-blue pencil). Repeat this for each triangle leg until all have been segmented according to the 10-foot contour interval.

It is now possible to sketch the contour lines, using a drafting pencil. Start with the highest contour, 370′ (figure 3.14H). Connect each 370′ point to form the contour line. Then proceed to the next lower contour (360′), and so on. The finished map (figure 3.14I) can be reproduced using a camera or copier (your nonphoto-blue lines will disappear, leaving a "clean" map).

Profiles

A profile is a convenient way to show the relative slope along a line of traverse. It is as if a knife cut vertically into the landscape along a

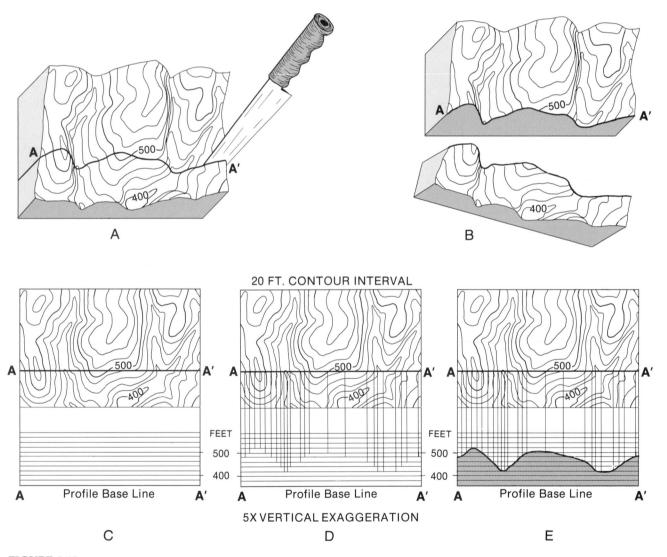

FIGURE 3.15
Cross-sectional profiles. (A) Illustrating the concept of a cross-sectional profile using a perspective terrain model with contour lines. (B) Profile AA′ revealed. (C) Layout of parallel elevation lines in relationship to profile line AA′. (D) Transferring elevation data to profile diagram. (E) Completed profile diagram.

traverse (figure 3.15A) and then the land was spread open and viewed from the side (figure 3.15B).

To create a profile is a reasonably simple task. Begin with a topographic map (figure 3.15C). Locate the position of the profile line (AA'). Draw a *profile base line* below the map and parallel to the profile line. Then draw a series of equally spaced and parallel elevation lines above the base line. (The spacing of the parallel lines is dependent on the contour interval, map scale, and the vertical exaggeration you desire.)

There are two scales that you must be concerned with when creating a profile diagram: The *horizontal scale*, which is the map scale, and the *vertical scale*, which is indicated by the contour interval. Under most conditions, the vertical variation in landscape is so slight in comparison with the horizontal scale that it is quite difficult to compare changing gradients of slope. Therefore, it is general practice to exaggerate the vertical scale when constructing a profile. Vertical exaggerations of 5X to 10X are quite common.

The spacing of the parallel elevation lines can be computed quite easily. For example, on a 1:24,000-scale map, one inch represents 2,000 feet on the ground. If the map had a 20 ft contour interval, the distance between the parallel elevation lines of the profile would be 0.01 of an inch! (This is computed by dividing the 20 ft contour interval by the number of feet per inch, based on the map scale—20 ft/2000 ft per inch = 0.01 inch.) Therefore, a 100-foot change in vertical elevation would be represented by 0.05 inch. At this scale it would be very difficult to discern relative change in slope. However, if the vertical scale is exaggerated, then slope change is more readily detected. In the example, a 10X (ten times) exaggeration would require 0.5 inch to represent 100 feet of vertical change.

Once the vertical exaggeration is decided, the parallel elevation lines of the profile can be constructed (figure 3.15C). After drawing the lines, assign them elevation values. Now the profile can be created.

Begin by locating points on the map where contour lines cross the profile line (figure 3.15D). From these points draw perpendicular

lines down to the corresponding elevation line on the profile diagram. When all points of intersection along the profile line are extended down to the profile diagram, connect the points to form the profile (figure 3.15E).

INFORMATION SHOWN ON MAP MARGINS

The space outside the neatline on a published USGS map contains information that identifies and explains the map. This marginal information corresponds somewhat to the table of contents and introduction of a book—it tells briefly how the map was made, where the quadrangle is located, what organizations are responsible for the contents, and gives other information to make the map more useful.

Map Identification

Each map is identified in the upper right margin by its quadrangle name, the state or states in which it is located, series, and type.

Quadrangle Name The name selected for a quadrangle is intended to identify the mapped area to the greatest number of people. Usually, each quadrangle is named for its principal city or place, or its most prominent feature, provided that all or most of the feature lies within the quadrangle. If that name designates another quadrangle of the same series in the same state, the name of a secondary place or feature is selected. Duplication of a quadrangle name within a state is permissible only when the name is for a feature common to maps of different series. For example, "Williamsburg" is an appropriate name for the 30-minute, 15-minute, and 7.5-minute Williamsburg, Virginia quadrangles.

State Name If the quadrangle includes areas of more than one state, the state names are shown in the title in order of decreasing area, even if only the part falling in one state is mapped.

County Name The name is not shown in the title if it appears within the mapped area.

Quadrangle Series and Type Series refers to the area mapped in terms of minutes or degrees (7.5-minute, 30 × 60-minute). The type is either topographic or planimetric.

Relative Position If a new 7.5-minute map covers a part of a published 15-minute quadrangle, a note is added giving the position of the 7.5-minute quadrangle; in the illustration, northeast quadrant (NE/4) of the Frostburg 15′ quadrangle.

CUMBERLAND QUADRANGLE
MARYLAND-PENNSYLVANIA-WEST VIRGINIA
7.5 MINUTE SERIES (TOPOGRAPHIC)
NE/4 FROSTBURG 15′ QUADRANGLE

Title Block The title block in the lower right margin shows the quadrangle and state name, the geographic index number, and, for 7.5-minute maps, the quadrangle's position in relation to the 15-minute map, if applicable. The title block also shows the year of the map. The geographic index number is the geographic position of the corner of the map nearest the Greenwich meridian and the equator, followed by the series, such as 7.5-minute.

CUMBERLAND, MD.–PA.–W.VA.
NE/4 FROSTBURG 15′ QUADRANGLE
N3937.5—W7845/7.5
1949

DMA 5263 III NE-SERIES V833

Cooperative Credit States and municipalities that cooperate in the preparation of USGS quadrangle maps by contributing funds are given appropriate credit on the published maps. The address line may be shown but is generally considered unnecessary. This heading appears in the center of the upper margin.

STATE OF GEORGIA
DEPARTMENT OF NATURAL RESOURCES
GEORGIA GEOLOGIC SURVEY

If the cooperator contributes to only a minor extent, the cooperation is shown only by the credit legend, which appears in the lower left margin. For example:

> Mapped, edited, and published by the Geological Survey
> in cooperation with State of Michigan agencies
> Control by USGS, USC&GS, and City of Detroit

Federal agencies that cooperate in preparing USGS quadrangle maps, either by contributing funds or by performing mapping operations that are incorporated in the published map, are listed only in the credit legend.

> Mapped by the Army Map Service
> Edited and published by the Geological Survey
> Control by USGS USC&GS, and USBR

The Department of the Interior–Geological Survey heading, in the upper left margin, is always the same in composition and placement.

UNITED STATES
DEPARTMENT OF THE INTERIOR
GEOLOGICAL SURVEY

Lower Margin Data

Magnetic Declination The magnetic declination for the year of field survey or revision is determined to the nearest 0.5 degree from the latest isogonic chart. It is shown by a diagram centered between the credit legend and bar scale.

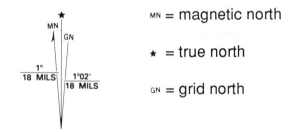

MN = magnetic north

★ = true north

GN = grid north

UTM GRID AND 1982 MAGNETIC NORTH
DECLINATION AT CENTER OF SHEET

Scales The center of the lower margin contains the following information, in the order indicated:

1. Representative fraction. Publication scale expressed as a ratio.

2. Bar scales in metric and customary units.

3. Contour-interval statement. If the map contains supplementary contours, a statement to that effect is added.

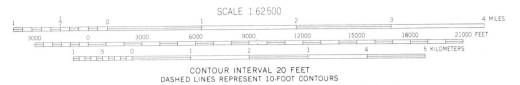

CONTOUR INTERVAL 20 FEET
DASHED LINES REPRESENT 10-FOOT CONTOURS

When the maximum elevation on the quadrangle is less than the specified contour interval, the following note is used in place of the standard contour-interval statement:

ENTIRE AREA BELOW 5 FEET

When a quadrangle area is located in two countries and the contour interval is different, the following note is used:

CONTOUR INTERVAL 20 FEET IN THE UNITED STATES
AND 10 METERS IN MEXICO

4. Vertical datum.
Prior to 1975:

DATUM IS MEAN SEA LEVEL

After 1975:

NATIONAL GEODETIC VERTICAL DATUM OF 1929

The statement "datum is mean sea level" is on standard USGS topographic maps printed prior to 1975. Maps printed since 1975 have the statement "National Geodetic Vertical Datum of 1929." The change was made to eliminate confusion arising from the difference between local "mean sea level" and the vertical control datum.

5. Depth-curve and soundings statement. Where applicable, "Datum is mean low water" is used for the Atlantic and Gulf coasts, and "Datum is mean lower low water" for the Pacific coast. The datum for the Great Lakes is low water to the nearest 0.1 foot as shown on the U.S. Lake Survey chart.

DATUM IS MEAN LOW WATER

DATUM IS MEAN LOWER LOW WATER

DEPTH CURVES AND SOUNDINGS IN FEET -- DATUM IS LOW WATER 576.8 FEET

6. Shoreline and tide-range statements:

SHORELINE SHOWN REPRESENTS THE APPROXIMATE LINE OF MEAN HIGH WATER
THE MEAN RANGE OF TIDE IS APPROXIMATELY 4 FEET

These statements are shown on maps that include tidal shorelines.

7. Map accuracy statement:

THIS MAP COMPLIES WITH NATIONAL MAP ACCURACY STANDARDS

The standard statement is, "This map complies with national map accuracy standards." Its absence means that, in some respects, the map may not comply with accuracy standards.

Road Symbols A legend explaining the various road symbols that are shown on the map is placed in the lower-right margin. This legend is tailored for each map to include only the classes of roads and the route markers that are shown in the mapped area. Trails are not included in the legend unless there are no roads on the map.

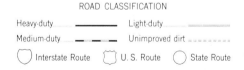

Quadrangle Location Map The location of the quadrangle within the state is shown by a diagram centered between the road legend and the bar scale. The small square representing the quadrangle mapped is positioned accurately. If a quadrangle falls within two or more states, the state having the largest area of quadrangle within its borders is shown (Maryland example). However, the new design of topographic maps employs a slightly different location system by showing several states or parts of states (Tennessee-Georgia example).

QUADRANGLE LOCATION

QUADRANGLE LOCATION

Adjoining Quadrangle Names Adjoining quadrangle names ("Cresaptown," "Patterson Creek") are shown so that map users may know that topographic data is available for these adjacent areas.

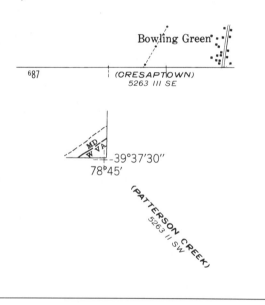

The following rules govern the use of these names:

1. Names of adjoining maps that appear on the current sales indexes, have been completed for publication, or will be completed in the near future, are shown on all four sides and all four corners.

2. Only the quadrangle name is shown if the adjoining map is published at the same scale or is of the same series.

3. Both the name and the scale are shown if the adjoining map is of a different series, as where a 15-minute map adjoins a 7.5-minute map, or two 7.5-minute maps adjoin a 15-minute map.

4. If there is map coverage of the same adjoining area in two series, one of which is the same as that of the map being prepared, the name of only the map of the same series is shown.

Projection and Grid Labels

Geographic Coordinates Geographic coordinates are shown at all four projection corners, and along the projection lines at 2.5-minute intervals for 7.5-minute maps, and at 5-minute intervals for 15-minute maps. When a map has an overedge area and one or more projection lines are extended, the coordinate value is positioned at the end of the extended projection lines. Coordinates for both the State Plane Coordinate and the Universal Transverse Mercator (UTM) grid systems are shown near the projection corners.

40'   694   SUFFIELD 1.3 MI.   INTERIOR—GEOLOGICAL SURVEY, RESTON, VIRGINIA—1979   42°00'
                              WINDSOR 11 MI.   630 000 FEET (CONN.)   72°37'30"
                                                              696000m E

State Plane Coordinate System Zone Grid If a quadrangle lies entirely within one grid zone, the numerical values of the grid lines are indicated for the x- and y-ticks nearest the southwest and northeast corners of the quadrangle. If a quadrangle lies in two or more grid zones, the second zone grid values are shown for the x- and y-ticks nearest the southeast and northwest corners; the third, at the southwest and northeast corners on the tick next to the first grid zone figures; and the fourth zone grid, in the southeast and northwest corners of the tick next to the second grid zone figures. Zone sequence is in order of decreasing area.

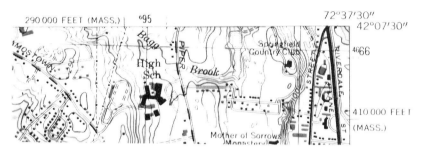

UTM Grid   In 1974, the USGS stated its intention to show a full fine-line UTM grid on its published maps at scales of 1:1,000,000 or larger, except in special cases for which it is not appropriate or justified. As of 1978, this intention had been implemented for new 1:25,000- and 1:100,000-scale line maps and 1:24,000-scale orthophotographic maps (an aerial photo base with superimposed topographic information). The UTM grid is in addition to, and does not replace, reference systems previously indicated.

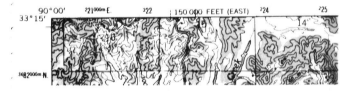

Other Marginal Data

Credit Legend   The credit legend is located in the lower-left margin. Because of the almost infinite number of possible combinations of data, credit legends cannot be rigid. The information is listed in the following order:

1. Name of the mapping agency.

2. Name of the editing and publishing agency. If one agency is responsible both for the mapping and for the editing and publishing, the two notes are combined.

3. Name of the agency or agencies that furnished the geodetic control.

4. Method or methods by which the mapping was performed.

5. Credit note for hydrographic information.

6. Informative and explanatory notes.

Produced by the United States Geological Survey
Control by USGS, NOS/NOAA, Tennessee Valley Authority, National Park Service, and Georgia Geodetic Survey

Compiled by photogrammetric methods from aerial photographs taken 1975.   Field checked 1975.   Map edited 1982
Supersedes Tennessee Valley Authority map dated 1969

Projection: Georgia coordinate system, west zone (transverse Mercator).   10,000-foot grid ticks based on Georgia coordinate system, west zone, and Tennessee coordinate system
1000-meter Universal Transverse Mercator grid, zone 16
1927 North American Datum.   To place on the predicted North American Datum 1983 move the projection lines 6 meters south and 4 meters west as shown by dashed corner ticks

Fine red dashed lines indicate selected fence and field lines where generally visible on aerial photographs.   This information is unchecked

Gray tint indicates areas in which only landmark buildings are shown

There may be private inholdings within the boundaries of the National or State reservations shown on this map

Year of Data   The year of the data shown on the map is printed beneath and as part of the title in the lower-right margin. The latest data of field completion or revision is used. It remains unchanged in future reprintings, but is changed when the map is revised.

Revisions shown in purple compiled in cooperation with State of Massachusetts agencies from aerial photographs taken 1977 and other source data.   This information not field checked.   Map edited 1979

WEST SPRINGFIELD, MASS.—CONN.
N 4200—W 7237.5/7.5

1958
PHOTOREVISED 1979
AMS 6468 II SW—SERIES V814

If minor cartographic corrections are made for a reprint, a note in small type is added under the year.

Publishing Agency   Federal printing and binding regulations require a note on all maps and other publications issued by government agencies showing the name of the department responsible for the publication, and the name and location of the printing plant. For USGS publications, the year of printing is included in the plant-imprint note.

The note is placed immediately below the border or the projection line in the lower-right margin, and may be moved to the left if necessary to avoid interference with other marginal data such as road destinations, grid ticks, and grid labels.

Road Destinations Road destinations are shown in the map margin for convenience in determining the distance to the next town or important road junction beyond the map border, and to facilitate orientation of the map with respect to well-known features.

Range and Township Numbers On 7.5-minute and 15-minute maps, range and township numbers normally are placed in pairs just outside the neatline. If one of a pair interferes with other marginal lettering, both numbers are moved to avoid the interference and maintain

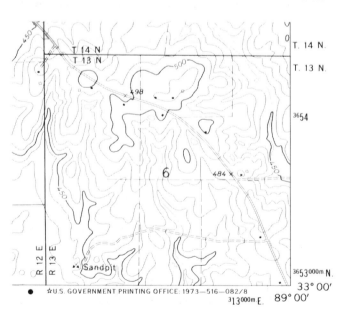

the equidistant spacing from the point where the range or township line intersects the projection line.

If a range or township line does not extend to the projection line, the numbers, without punctuation, are placed inside the map near the end of and straddling the line.

Available Larger-Scale Coverage A statement identifying the available larger-scale coverage is placed above the lower-margin title.

If the area of a 15-minute map being reprinted has coverage by one or more 7.5-minute maps, a statement patterned after the one shown is placed below the road-classification legend.

> This area also covered by 7.5-minute, 1:24 000-scale maps: West 1964, Bowling Green 1964, Owens Wells 1964, and Durant 1964

> This area is also partially covered by 1:24 000 scale maps of Olds Ferry NW, Olds Ferry, and Olds Ferry SE 7.5 minute quadrangles

Shaded-Relief Edition A map for which there is a shaded-relief edition carries the statement, "This map is available with or without shaded relief overprint." The statement is placed in the center of the lower margin, directly above the map-accuracy statement.

THIS MAP IS AVAILABLE WITH OR WITHOUT SHADED RELIEF OVERPRINT

MAP REVISION

Maps are a static "picture" of dynamic geographic conditions, so map revision is inevitable. How much change must occur before a map is a candidate for revision? More specifically, what kind of revision is necessary—total revision, partial revision, photorevision, or photoinspection?

Total Revision

A phased or sequential approach is usually employed in total revision. It yields this sequence of map products:

1. Orthophotoquad. An orthophotoquad is prepared from high-altitude, quad-centered photographs. At first glance, an orthophotoquad often resembles a photo more than a

map. On closer inspection, you can see drafted roads, contour lines, and other commonly used topographic symbols. These maps may be available in printed, photographic, or diazo ("blackline") form.

2. Advance planimetric edition. An advance planimetric edition, lacking topographic contours, generally precedes the final topographic map. It is prepared by direct scribing on an orthophotoquad base. In compiling this edition, some stereoscopic compilation from aerial photos may be required for accurate positioning of obscure features, such as single-line drains or drains in timbered areas. Nonphotographic source materials are used when available to add boundaries, names, and other features not visible on the photographs. This product is available in black-and-white printed or diazo form.

3. Totally revised map. In the total-revision process, the advance planimetric edition is field-checked, and additional field information is obtained for completing the map. Contours and additional planimetric features are stereocompiled. The planimetric additions and corrections are then applied to the color-separation materials produced for the advance planimetric edition. Parts of the old map may be used if they are known to be of standard quality and accuracy. Content of the total revised map complies with current specifications for new maps.

Partial Revision

Procedures for partial revision vary with the deficiencies to be corrected. Corrections may range from minor updating—adding a single reservoir or highway—to field-checking of revised features. Aerial photographs are used in stereoscopic or monoscopic mode, as appropriate. Surveys, engineering plans, and other source materials are also used when available. Changes made by partial revision are shown on the published map in conventional colors.

Photorevision

Corrections, additions, and deletions of planimetric features are made by photointerpretation and transfer of detail from aerial photographs. Either monoscopic or stereoscopic procedures are used, depending on terrain relief. Other available source materials, such as surveys, engineering plans, or local maps, are used to revise boundaries, names, or other features not visible on the photographs. The revised features are not field-checked, and the new information is printed in a bright purple color on the new map.

Photoinspection

Quadrangles are selected for photoinspection in response to commitments to statewide cyclic inspection programs or other requirements. Areas known to have a high rate of change, such as urban-growth areas, are generally reviewed on a five-year cycle. Generally, maps are reviewed every five to twenty years.

Photoinspection criteria used to determine the need for revision are given in figure 3.16. Three examples of inspection results that qualify a quadrangle for consideration for total revision are:

1. Total change of linear features (highways, streams, pipelines) exceeds 50 miles.

2. Total change of areal features exceeds 15 square miles or 25% of the land area.

3. Total change of linear and areal features equals or exceeds individual linear or areal criteria. For example, 30 miles of changed linear features amount to 60% of 50 miles, and 6 square miles of areal change is 40% of 15 square miles. Together the 60% and 40% change comprise the full amount of change necessary for revision consideration.

The following factors, along with the criteria in figure 3.16, are considered in determining the type of revision:

1. Cartographic materials. Maps for which cartographic materials (color-separation negatives or positives, or manuscripts) are nonexistent, or in poor condition, are candidates for total revision.

2. Amount of change. Maps of good geometric quality, but which need updating, receive

Revision Criteria for Topographic Quadrangle Maps Determined by Photoinspection

	Major changes[1]	
Feature	*Description*	*Minimum change required for revision*
Roads, major	Interstate and major highways of at least four lanes.	1 mile per quad (or in lesser amounts if necessary to preserve continuity of the feature through a block of several quadrangles).
Roads, minor	Double-line roads symbolized by 40-foot road width.	5 miles per quad with no segment less than 1 mile.
Drainage	Double line. Single line, perennial	1 mile per quad. 5 miles per quad with no segment less than 1 mile.
Reservoirs	Water storage, controlled outlet.	1 mile or more in length and covering at least 0.25 mi².
Airports	Major landing fields: private, commercial, or military, generally hard surfaced.	New runways, additions, or changes of at least 0.5 mile.
Strip mines	Areas of disturbed earth, active or inactive.	Extensions of mining area or reclaimed areas of at least 0.25 mi².
Urban-suburban	Areas in and surrounding metropolitan areas.	Subdivision-type pattern of streets and buildings covering at least 0.125 mi².

	Minor changes[1]	
Feature	*Description*	*Minimum of three changes or more to justify revision*
Ponds	Stock tanks, private ponds.	Eight per quad with average width 200–300 feet.
Airfields	Private landing fields, generally not hard surfaced.	One or more, regardless of size.
Industrial sites	Large areas, usually with rail or highway access, designated as industrial parks, manufacturing, or commercial areas.	One or more buildings totaling 100,000 ft² or more.
Woodland	Major timber areas or orchards.	Total of 2 mi² of addition or deletion with no area less than 1 mi².
Miscellany	Pipelines, major power transmission lines, ditches.	5 miles per quad with no segment less than 1 mile.

	"Total revision" criteria[2]	
Feature	*Description*	*Minimum change required for total revision*
Linear	All roads, railroads, drainage, coastline, and airport runways and taxiways.	50 miles.
Areal	Reservoirs, strip mines, urban-suburban, and beaches.	15 mi² or 25 percent of land area.

[1] When none of the major changes or fewer than three of the minor changes occur on a map being reviewed, the photoinspection date and note will be added to reprint editions.

[2] Features will not be double counted. For example: (1) if a double-line stream change is counted as a linear change, the open-water tint will not be counted as an areal change, or (2) if the urban-tint change is counted as an areal change, minor roads in urban-tint areas will not be counted as linear changes.

When reviewing maps, the cumulative change that would be shown in purple overprint with this anticipated revision is considered. Any map on which the summation of both categories approaches 100 percent will be considered for total revision rather than photorevision.

FIGURE 3.16

Revision criteria for topographic quadrangle maps, as determined by photoinspection. (from Thompson 1979)

(a) total revision if the planimetric change is extensive; (b) partial revision if the planimetric changes are not extensive, but the map warrants contour correction or classification verification (roads, buildings, boundaries); and (c) photorevision if the planimetric change is moderate.

3. Contour interval. Maps with an interval two or more steps larger than designated in the current interval plan are candidates for total revision.

4. Accuracy. Maps which fail to meet established accuracy standards are candidates for total revision.

5. Survey date. Maps can be separated into three groups of implied accuracy, based on methods used during the original compilation. These methods can in turn be related to dates. Generally, (a) maps prepared prior to 1948 do not meet established horizontal-accuracy standards; (b) maps surveyed between 1948 and 1957 have questionable horizontal accuracy and should be checked; and (c) maps surveyed after 1958 have acceptable horizontal accuracy and need to be checked only if reasonable doubt exists.

In evaluating a map to select the appropriate revision procedure, evaluation of adjoining maps is also considered.

The review date and an explanatory footnote are added to reprint editions of those maps which, upon comparison with recent source material, are found to require no revision. If a map has been inspected more than once and found to be adequate, only the most-recent review date is added.

Depending upon the availability of bathymetric data, all coastal quadrangles are converted to topographic-bathymetric maps at the time of revision.

CAMEO

Morris M. Thompson
United States Geological Survey (Retired)

Morris M. Thompson received a B.S. degree in Engineering in 1934 and a B.S. in Civil Engineering in 1935 from Princeton University. From 1935 through 1938, he worked in upstate New York as a cost and progress engineer with the U.S. Soil Conservation Service. In 1939, he began his career in the national mapping program of the U.S. Geological Survey, which continued until his retirement in 1975.

A specialist in topographic mapping by photogrammetric methods, Mr. Thompson advanced through a series of increasingly responsible positions: Photogrammetric Unit Supervisor in the Tennessee Valley mapping project at Chattanooga; Chief, Photogrammetric Section, Atlantic Region, Arlington, Va; Atlantic Region Engineer; Assistant Chief Topographic Engineer, Washington, DC; Chief, Office of Research and Technical Standards, Reston, Va. He was instrumental in introducing modern methods in USGS map production.

Author of more than 50 articles published in various scientific journals, Mr. Thompson is perhaps best known as the author of *Maps for America*, published by the USGS in 1979 and now in its third edition (1988). He also served as the editor-in-chief for the third edition of the *Manual of Photogrammetry*, published by the American Society of Photogrammetry.

Mr. Thompson has served with distinction as an Executive Committee member of the Surveying and Mapping Division, American Society of Civil Engineers. He has served as the Chairman of the Cartography Division, American Congress on Surveying and Mapping. Furthermore, he served as Chairman of the Commission on Automated Cartography of the International Cartography Association, 1968–72.

The recipient of numerous awards, Mr. Thompson holds three major American Society of Photogrammetry awards: Talbert Abrams Award for authorship (1950), Fairchild Award for achievement in photogrammetry (1966), and Honor Member Award for eminent career (1978). Additionally, he received the ASCE's Surveying and Mapping Award (1978) and the ACSM's Cartography Honors Award (1980).

Since his retirement, Mr. Thompson has remained active on a part-time basis with USGS, as well as managing a consulting business.

Introduction

With the burgeoning of new needs for maps, and the development of advanced cartographic technology to meet those needs, different kinds of maps are being produced in ever-broadening variety. Yet the topographic map remains as the old standby—the map with the most complete information about the land and works of man upon that land, and the map which serves as the base for an infinite spectrum of other maps, each made for some special purpose.

The topographic map contains the information that enables engineers to pinpoint the best site for a dam and to locate the shoreline of the lake that will form behind the dam. It tells construction contractors how much earth will have to be moved to build the new highway they have laid out on the topographic map. It enables builders to select the optimum layouts for streets and houses in a new development. It shows communications specialists the best locations for microwave relay stations. It tells foresters how wooded areas are related to elevation and slope of terrain. It gives land-use planners information on how the land should be utilized for maximum benefit to the environment and the economy. It enables vacationers to select the best hiking trails, fishing spots, and points of interest. In short, if one needs accurate information on the roads, railroads, airports, docks, streams, ponds, bridges, hills, valleys, woodlands, parks, or political boundaries in a given area, the most complete source is the topographic map.

Elements of a Topographic Map

In general, a topographic map is composed of these elements: control, culture, hydrography, hypsography, vegetation, boundaries, place names, and marginal information.

A network of points of established positions and elevations forms the framework of the map. This network, called "control," provides the means of tying features on the map to their proper positions on the Earth's surface. Control points are established by precise geodetic surveys, and are shown on the map in black or brown.

The works of man, shown on a topographic map in black, are known as "culture." Included in culture are roads, streets, trails, railroads, airports, dams, bridges, tunnels, buildings, mines, powerlines, pipelines, industrial areas, cemeteries, recreation areas, and other manmade features.

The water features of a topographic map are called "hydrography." These features, shown in blue, include rivers, creeks, branches, natural lakes and ponds, manmade lakes and reservoirs, canals, aqueducts, landmark springs and wells, waterworks, beaches, bays, inlets, estuaries, and other coastal features.

Map features depicting relief of terrain are known as "hypsography." On topographic maps, the approximate elevation of any point can be read directly or interpolated from contours. The hypsography not only permits the map user to determine measurements of hills, valleys, and plains, but it also presents a graphic picture of the ground surface.

Vegetation, depicted in green on topographic maps, includes woodland, scrub, orchards, vineyards, mangrove, and wooded marsh. It is shown if the size and density of a vegetated area meets certain criteria.

Civil boundaries shown on topographic maps are the limiting lines of jurisdictional authority for countries (international), states, counties, cities, incorporated places, national and state parks, and certain other units of civil government. Boundary lines are shown in black, with a different linear symbol for each category of boundary.

The names of places and features shown on topographic maps are those in local usage, as nearly as can be determined. The most important names are included as appropriate for the scale of the map.

Marginal information is shown in the space outside the neatline of the topographic map. It tells briefly how the map was made, where the area is located, what organizations provided the content, and gives more information to make the map more useful.

Topographic Mapping Operations

To use a topographic map intelligently, you should certainly be aware of the map elements mentioned above. Furthermore, you can enhance your ability to obtain maximum information from the map by having a general understanding of the steps involved in making the map.

In the 1980s, as this is being written, the procedures for mapmaking are undergoing a revolution in technology. Some of the operations that are considered to be "conventional" methods are being replaced by "high-tech" procedures utilizing computers, satellite-borne sensors, and automated cartographic equipment. But, since most of the topographic maps now available were produced by conventional methods, it is important to understand these methods.

In general, the conventional topographic-mapping system comprises these phases: planning, aerial photography, control surveys, photogrammetric compilation, map finishing, reproduction, and distribution. An examination of this sequence should make it clear that mapmaking is principally an information-gathering task, quite different from the prevalent concept in the public mind that someone simply sits down at a drafting table and "draws a map."

The first step in the mapping sequence is to plan the mapping project. Where is the area to be mapped? What are its boundaries? What kind of map is needed? How will the project be funded? What procedures should be used? What manpower and equipment will be needed? How should operations be scheduled? These and other questions must be answered by the map-production planning group, usually consisting of professional engineers and cartographers.

Most of the map detail for modern topographic mapping is obtained from aerial photography. The map-producing organization develops the specifications for the photography and usually awards a contract to a company specializing in the extremely precise aerial photography required for mapping. Highly trained pilots and aerial photographers perform this operation, making sure that stereoscopic photographic coverage is obtained for every square foot of the project area.

After the aerial photography is delivered to the mapping organization, photogrammetrists plan the control for the mapping project, marking identifiable points on the photographs for which positions or elevations are desired. The photos are then taken to the project area by field engineers and technicians for obtaining the required elevations and positions by means of surveying instruments. The established positions are plotted on the map base.

Once the control is available, photogrammetrists can orient the photographs correctly in the plotting machine and proceed to compile the map detail on the map base. The usual compilation sequence is: major hydrographic features, culture, hypsography, minor hydrography, and vegetation.

Because of dense vegetation and other obstructions, the photogrammetrist may not be able to plot all the detail from the aerial photographs. If necessary, engineers take copies of the partly completed map to the project area for field completion. At this time, they also obtain the names of places and features included in the mapped area.

The map is now ready for map finishing. Cartographers and cartographic technicians perform smooth drafting as necessary, add names, and edit the finished map to make sure there are no errors. A separate sheet is produced for each color in which map detail is to be printed.

The color-separated sheets are delivered to the reproduction unit, which has multicolor printing presses. The maps are printed from plates corresponding to each color separate, to produce the final multicolored maps. The maps are now ready for distribution to the map users.

In this outline of topographic mapping operations, reference has been made to cartographers, engineers, photogrammetrists, and technicians. Although individuals perform different tasks, as indicated by these designations, all of them function within the field of cartography, as they have the common objective of producing maps. In U.S. Government practice, many of these individuals have the title of "cartographer" or "cartographic technician." [Increasingly, the titles of "physical scientist" and "computer scientist-analyst" are employed to identify personnel involved in mapmaking.—The Authors]

New Trends in Topographic Mapping

In 1975, this writer predicted that by the year 2000 the classical form of maps in which features are represented by lines and symbols would be supplemented or replaced to a large extent by photo-image maps or sensor-image maps (see M. M. Thompson, "Surveying and Mapping in the Year 2000." *Proceedings of the 1975 Convention,* American Congress on Surveying and Mapping). Cartographic information would be compiled and stored in digital form in central data banks, and, by means of computer access to this information, any desired kind of map could be produced, or an old map updated by computer-graphics techniques. The principal data-gathering machinery for much mapping would be spacecraft carrying arrays of high-resolution sensors. The prediction concluded that remote sensing, satellites, and computers would be the future of cartography.

As of 1986, this prediction has, to a considerable extent, already been fulfilled. A National Cartographic Data Base has been established at the U.S. Geological Survey with the object of providing a wide range of data in computer-compatible form. The initial categories of data are digital elevation models, public land net, boundaries, hydrography, and transportation. Automated systems are operational for simultaneous production of orthophotographs, contours, and digital terrain models. By manipulation of images from various sensors, systems have been developed for theme extraction and derivation of specialized cartographic information. The development of a completely automated system is in an advancing stage.

[It is also desirable for agencies to develop an international cartographic data base. To some degree this has been accomplished by the Central Intelligence Agency (CIA) with its small-scale World Data Base I and II. Larger-scale data bases represent a more complex compilation problem. In part, the problem originates from the fact of inconsistent and incomplete mapping of the world. Once a larger-scale data base has been acquired, it must be shared. In fact, strategic weapons surveillance will not be possible without it.—The Authors]

Essentially, these new trends in cartography have progressed because of applications envisioned in topographic mapping. But there are countless spin-offs of topographic-mapping advances to other cartographic applications. The same high technology that works for topographic mapping can be used for producing cadastral maps, thematic maps, census maps, land-use maps, geologic maps, economic maps, political maps, weather maps, hydrographic and aeronautical charts, and so on. If all the topographic data for an area has been digitized, it provides a base for at least a good start on each of the other products named above.

You might ask, "If cartography is changing so much, why do I have to study the methods of the past? Why can't I just skip to the new technology?" The answer is, of course, that you must learn to walk before you can run. You must know something about magnetism before you can understand electricity, and you must understand electricity before you can master electronics. This chapter on topographic maps deals with the graphical presentation of these maps, and the lines and symbols will remain the same, whether the map is produced by conventional methods or by high technology. If, in the future, new computer-compatible graphic symbols are established, or indeed if the map exists only in a data bank, you will have something new to learn.

SELECTED READINGS

Biddle, D.S., A.K. Milne, and D.A. Shurtle. 1974. *The language of topographic maps*. Milton, Australia: Jacaranda Press.

Bies, John, and Robert A. Long. 1983. *Mapping and topographic drafting*. Cincinnati: South-Western Publishing Co.

Chevrier, Emile D., and D.F.W. Aitkens. 1970. *Topographic and map and air photo interpretation*. Toronto: Macmillan Co.

Larsgaard, Mary L. 1984. *Topographic mapping of the Americas, Australia, and New Zealand*. Littleton, CO: Libraries Unlimited, Inc.

Knowles, R., and P.W.E. Stowe. 1976. *North America in maps: Topographical map studies of Canada and the USA*. London: Longman Group.

Raitz, R., and John Fraser Hart. 1975. *Cultural geography on topographical maps*. New York: John Wiley & Sons.

Thompson, Morris M. 1979. *Maps for America: Cartographic products of the U.S. Geological Survey and others*. Washington: U.S. Government Printing Office.

Tyner, J.A. 1973. *The world of maps and mapping*. New York: McGraw-Hill Book Co.

PROJECTS

PROJECT 3A

TOPOGRAPHIC MAP INTERPRETATION

Objective

This exercise will familiarize you with some of the more frequently used symbols and marginal information found on topographic maps. You will be gleaning facts (both direct and implied) from a sample section of a topographic map.

Materials and Equipment

1. Engineer's scale
2. Straight edge
3. Pencil
4. Figure 3.17 (topographic map symbol sheet)
5. Figure 3.23 (color topographic map, inside back cover)

Procedure

Answer the following questions by interpreting the portion of a topographic map shown in figure 3.23 at the end of the book.

1. What is the name of the topographic sheet? _Fib. 3.23_ _____

   _____

2. From what series is the sheet? _____

3. Portions of what states are on the sheet? _____

   _____

4. What is the horizontal scale of the map? 1: _____

5. What is the contour interval? _____ 20 FT _____

6. What are the geographical coordinates of the map at

   SE Corner   lat. _39° 37' 30'_   long. __78°45'___

7. List all of the agencies that contributed to the production of the map.

   _____

   _____

   _____

8. What quadrangles are adjacent to the map?

   S._____

   E._____

   SE._____

9. In what year was the declination of magnetic north determined for this map? _____

10. What is the magnetic declination for the map? _____

11. When were the aerial photos taken that were used to produce the map? _____

12. What is the most recent edit date for the map? _____

13. What grid systems are shown on the map?

    a. _____   b. _____

    c. _____   d. _____

14. List all of the different road symbols shown on the map.

    a. _____   b. _____

    c. _____   d. _____

    e. _____   f. _____

    g. _____   h. _____

15. List all of the other line symbols found on the map.

    Boundaries: a. _____   b. _____

                   c. _____   d. _____

    Physical: a. _____   b. _____

    Cultural: a. _____   b. _____

                   c. _____   d. _____

                   e. _____   f. _____

16. List all of the point symbols found on the map.

    a. _____   b. _____

    c. _____   d. _____

    e. _____   f. _____

    g. _____   h. _____

    i. _____   j. _____

    k. _____   l. _____

17. What are the highest and lowest points of elevation on the map?

    Highest _____   Lowest _____

PROJECT 3B

CROSS-SECTIONAL PROFILES FROM TOPOGRAPHIC MAPS

Objective

This exercise will show you how to construct cross-sectional profiles from topographic maps. It will illustrate the value of vertical exaggeration when depicting relief using cross-sectional profiles.

Materials and Equipment

1. Engineer's scale
2. Straight edge
3. Dividers
4. Pencil
5. Figure 3.18 (profile sheet)

Procedure

1. Construct two profiles along line AA' in figure 3.18. Construct the first profile without any vertical exaggeration. Construct the second profile with 5X vertical exaggeration. (Note which is easier to use when identifying variations in slope.)

2. Construct a profile found along line BB'. Use 5X vertical exaggeration. Based on this profile, if you were standing at point X, which of the following points could you see? (q, r, s, t, u, v)

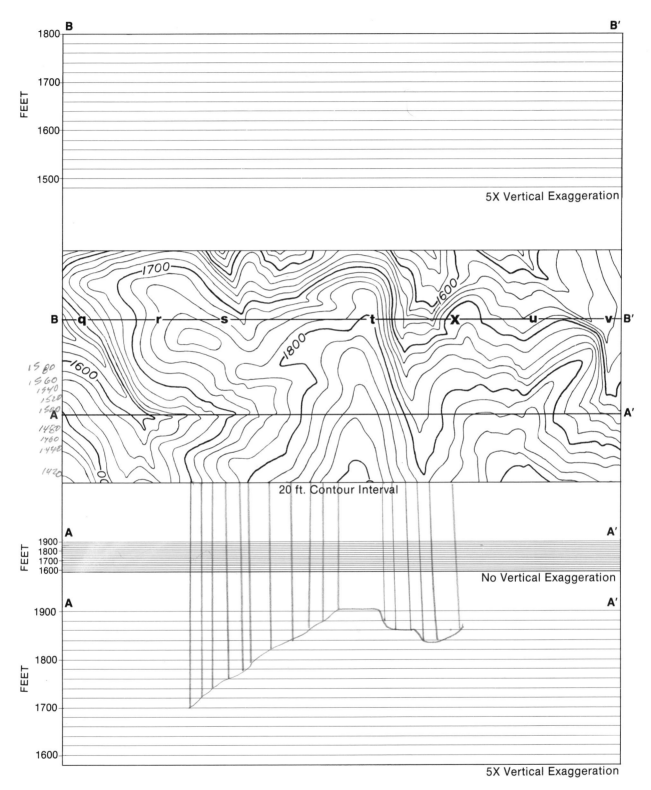

FIGURE 3.18
Profile sheet for Project 3B.

PROJECT 3C   **LANDSCAPE VISUALIZATION FROM TOPOGRAPHIC MAPS**

Objective

This exercise requires you to extract a visual image by interpreting information from a topographic map. You will learn to relate scale, position on the map, and information conveyed by topographic symbols, and use this information to construct a visual image of the landscape.

Materials and Equipment

1. Engineer's scale
2. Protractor
3. Straight edge
4. Dividers
5. Pencil
6. Figure 3.17 (topographic map symbol sheet, inside back cover)
7. Figure 3.23 (color topographic map, inside back cover)

Procedure

Assume you are hiking to points A, B, and C shown on the topographic map in figure 3.23. Describe verbally what you would see from points A, B, and C. Describe what you would see at each of these points when facing north, east, south, and west. Remember that you are standing on the ground, and certain features will be obstructed by other features. You are to describe only what you can actually see from each point of observation.

PROJECT 3D   **MAP REVISION**

Objective

This project will familiarize you with the procedure for determining whether a topographic map needs revision. You will do so by comparing a topographic map (figure 3.19) with a more recent aerial photo (figure 3.20). You will assess the differences you discover between the map and photo, and determine whether sufficient modification has transpired to merit map revision. If map revision is necessary, you must determine what level of revision is to be implemented.

Materials and Equipment

1. Engineer's scale
2. Drafting mylar or high-quality tracing paper. A transparency of the topographic map (figure 3.19) made on a copier at 100% size can be substituted for the drafting mylar.
3. Map measuring wheel or digitizer possessing a linear measurement function. (Note: If this equipment is not available, it is possible to use the sum of a series of straight lines that approximate a curved line to estimate its length.)
4. Polar planimeter or digitizer possessing an area-measurement function. (Note: If this equipment is not available, a simple dot grid can be used for reasonable approximation of area measurement.)
5. Permanent marker pens—
   Light green, broad tip
   Red, orange, black, and dark green, fine tip
6. Figure 3.19 (topographic sheet)
7. Figure 3.20 (aerial photograph)

Procedure

(Skip steps 1 and 2 if a transparency of the topographic map is used.)

1. Tape the drafting mylar over the topographic map. Using the black marker pen, trace all existing roads.

2. Using the light-green marker pen, delineate all woodland (the light-gray tint on the topographic map).

3. Align the mylar sheet (or transparency) over the aerial photograph. Using the red marker pen, trace all major roads (interstates and highways of at least four lanes) that do not appear on the topographic map.

4. Using the orange marker pen, trace all minor roads not appearing on the topographic map.

5. Compare woodland cover shown on the topographic map and the aerial photograph. Using the dark-green marker pen, note all changes (additions or deletions).

6. The map and photo scales are 1:24,000. Using the measuring wheel or digitizer, calculate the number of miles of major highway additions. Do the same for minor highway additions.

7. Using the polar planimeter or digitizer, compute the number of square miles of woodland change.

8. Assume that the topographic map used in this exercise is representative of the entire quadrangle, even though it covers only 20% of the area. Multiply by 5 all of the measurements, to represent the entire topographic quadrangle.

9. Compare your results with the criteria outlined in figure 3.16 to determine what kind of revision is necessary.

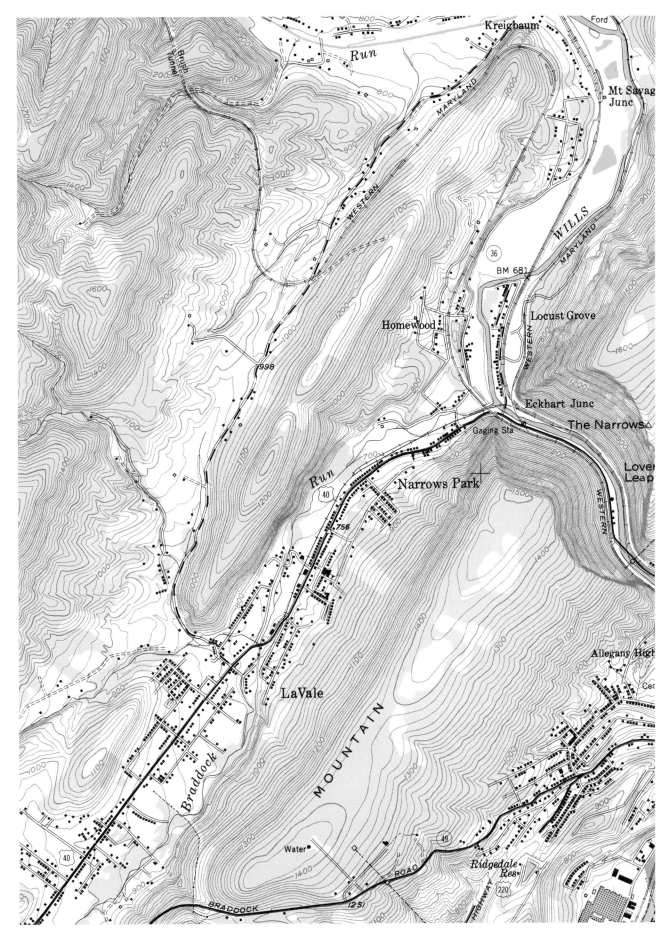

FIGURE 3.19
Topographic map for use in Project 3D. Scale 1:24,000.
(USGS)

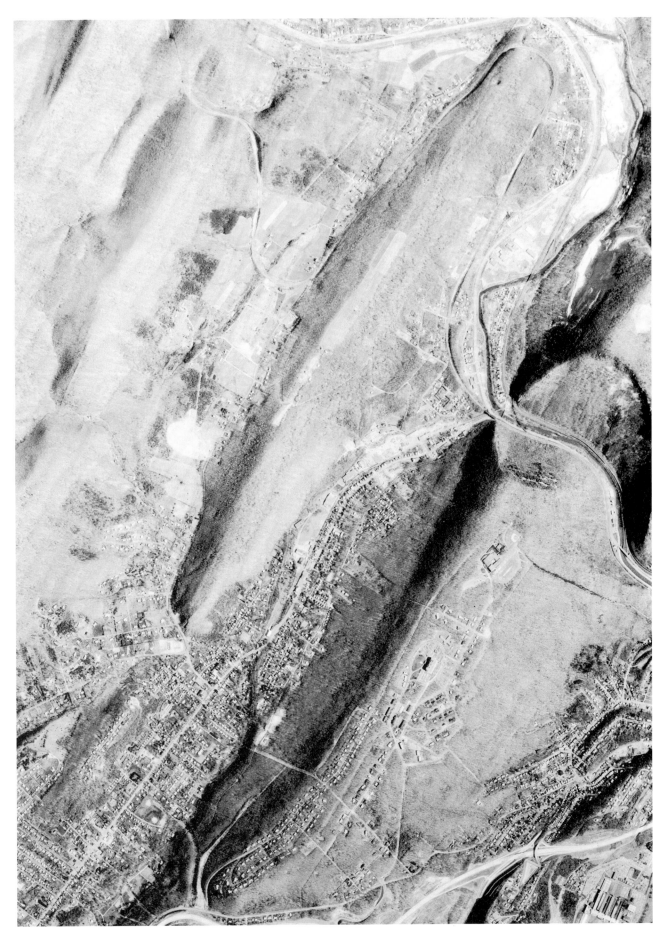

FIGURE 3.20
Aerial photo for use in Project 3D. Scale 1:24,000. (USGS)

PROJECT 3E

CONTOUR INTERPOLATION

Objective

This exercise will give you practice in simple linear interpolation to establish the position of contour lines from spot elevations.

Materials and Equipment

1. Drafting pencil and eraser
2. Nonphoto-blue pencil (such as Berol 919)
3. Set of colored pencils (optional)
4. Light table
5. Figure 3.21 (linear divider grid)
6. Figure 3.22 (topographic spot elevation sheet)

Procedure

Using the interpolation techniques discussed in the part of this section titled "Contour Interpolation," complete the contour map from the spot elevations in figure 3.22.

1. Begin by interpolating the intermediate elevations between adjacent points, using the linear divider grid (figure 3.21). Keep in mind the rule pertaining to interpolation when points are separated by streams.

2. Once interpolation is completed, connect points of equal elevation to form contours.

3. Optional—When all the contour lines have been completed, use colored pencils to add hypsometric tints:
   a. 1440'–1560' = dark green
   b. 1560'–1680' = light green
   c. 1680'–1800' = yellow
   d. 1800'–1920' = orange
   e. 1920'–2040' = red
   f. 2040'–2160' = brown

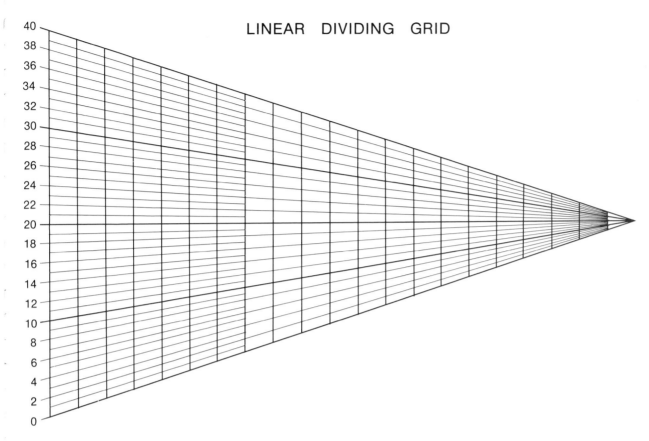

FIGURE 3.21
Linear dividing grid to be used for contour interpolation in Project 3E.

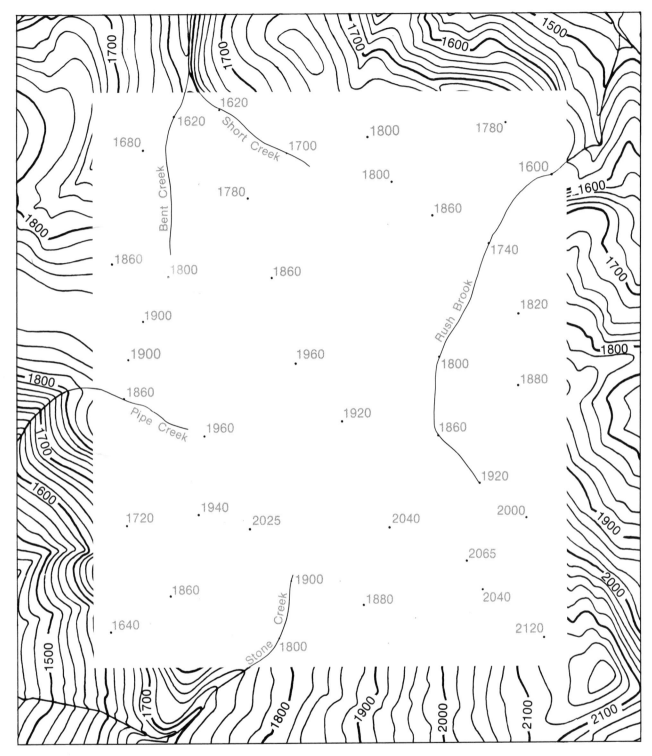

FIGURE 3.22
Partially completed contour map used in Project 3E.

INDEX

Accuracy criteria, 111
Acquisition of satellite images, 18–19
Advanced planimetric edition, 109
Aerial photographs, 1, 7
    forensic use, 11
    obtaining, 11
Aerial photography, 114
    determining availability, 49
    private sources, 13, 15
Aerial Photography Summary Record (APSRS), 12–13,
    16–17, 49
Alphanumeric index, 10
Alphanumeric system, township-range blocks, 85
Androit, John L., 31
Anglo-American cataloging rules, 43, 48
Annotation, LANDSAT image, 15–16, 18, 55
APSRS, 49
Atlases, historical, 31
Atlases, Library of Congress Collection, 23
Availability of source materials, 5
Azimuth, 58

Base line, 62, 85
Base maps, 1, 4, 29–31
Base material, 1
Birdsall, Stephen S., 31
Bow compass, 70
Bubble level, 62
Bureau of Census, 31
Burn, Christian F., 29

Carrington, David K., 22
Cartesian system, 94
Cartobibliographies, 22, 28–29
Cartographers, 114
Cartographic and Architectural Branch, National
    Archives, 7, 23, 28
Cartographic materials, National Union Catalog, 7
Cartographic technicians, 114
Central Intelligence Agency, 115
Chart, map symbol, 98
Civil War maps, 23, 26, 28
Cobb, David A., 29
Codes, Aerial Photography Summary Record, 17
Collecting field data, 75
Collections, 1, 5, 6
Color-separated sheets, 114
Comparative coverage quadrangle, 84
Compass, magnetic, 62
Compass bearing, 58
Compass-traverse maps, 58–61
Compass traverse reconstruction, 73
Compilation, map, 22
Compilation process, 4
Computer-Compatible Tapes (CCT), 18
Computer scientist-analyst, 114
Construction of a map, 1
Contour interpolation, 99–102, 133–37
Contour interval, 101
    revision criteria, 111
    statement, 105
Contour line, 9, 98

Control, 113
Cooperative credit, 104
Copyright, 29
    problems, 22
County boundaries, 31
County outline maps, 30
Creating maps, 3–4
Credit legend, 107
Cross-sectional profiles, 102, 121–23
Culture, 113
Cylindrical projection system, 88

Data, 1, 2
Data base, 6
Datum, 105
Day, James, 29
Defense Mapping Agency (DMA), 3
Dempsey, Patrick R., 23
Depth-curve, 105
Determining availability of aerial photography, 49
Dubriel, Lorraine, 22

Earle, Carville V., 31
Easting, 90
Ehrenberg, Ralph, 45
EOSAT, 10, 21

False origin of zone, 89, 90
Field mapping, 2, 57–63
Field notes recording sheet, 67, 69, 77
Field sketch map, 58, 74
Field survey, 1
Finding map collections, 22–23
Fire insurance maps, 23, 24–25
Florin, John W., 31
Foot made maps, 57–58, 65–66
    reconstruction, 71

Gap, 60
Geographic coordinates, 6, 10, 106
Geographic name lists, 10
Geography and Map Division, Library of Congress, 23
Geometric correction, 15
Grade, 99
Grid
    linear dividing, 135
    map, 84–94
    SPCS, 92–96
    USPLS, 85–87
    UTM, 86, 88–92, 107
Grim, Ronald E., 7, 22, 29, 31
Ground elevations, 99
Ground resolution, 12

Hebert, John E., 23
Historical atlases, 31
Horizontal scale, 103
Hydrography, 113
Hypsography, 113
Hypsometric information, 98

Identifying source maps, 35, 41
Image
    center, 15
    date, 15
    identification number, 18
Index
    alphanumeric, 10
    maps, 5, 6, 10, 12–13
    sheets, 10
Indexing system, LANDSAT, 21
Intermediate-scale maps, 83
Interpolation, 99–102
    contour, 133–137
    topographic map, 117–19

Karrow Jr., Robert W., 29
Kelsay, Laura E., 28

Landmark, 98
Land ownership maps, 23, 27
LANDSAT, 15
    image annotation, 55
    indexing system, 21
    nominal scene, 19, 52
    ordering, 19
    path and row numbers, 19, 52
LANDSAT-EROS, 10
Landscape visualization, 125
Land subdivision, USPLS, 85–87
Land-use mapping, 1
Larger-scale coverage, 108
Large-scale maps, 1, 3, 83
Latitude, 5, 6
LeGear, Clara Egli, 23
Legend, Credit, 107
Legibility requirement, 97
Library of Congress, 7, 10
    map collection, 23–28
    map reading room, 8
Line, contour, 9
Linear divider grid, 101, 135
Linear features, 98
Local anomalies, 59
Locating maps, 5, 29–31
Location, map, 5
Long, John H., 31
Longitude, 5, 6

Magnetic compass, 62
Magnetic declination, 59, 104
Magnetic north, 59
Manuscript maps, 7
Map
    accuracy statement, 105
    annotation printout, 44
    base, 1, 4, 29–31
    cataloging, 43–48
    cataloging form, 47
    Civil War, 23, 26, 28

Map, *continued*
  collections, 7
    finding, 22–23
    Library of Congress, 23–28
  compass-traverse, 58–61
  compilation, 2, 22
  construction, 1
  county outline, 30
  creating, 3–4
  elements, topographic, 113
  field sketch, 74
  fire-insurance, 23, 24–25
  foot-made, 57–58
  grids, 84–94
  identification, 103–6
  index, 6, 10
  information, marginal, 103–8
  intermediate-scale, 83
  inventory sheet, 39
  land ownership, 23, 27
  large-scale, 1, 3, 83
  locating, 5
  panoramic, 23
  plane-table, 62–63, 79–82
  published, 5
  quadrangle location, 105
  reading room, Library of Congress, 8
  reference, 3
  reference code, 14
  revision, 108–11, 127–31
  Revolutionary War, 29
  scale, 6
  series, 4, 94–95
  small-scale, 83
  source, 5
    national, 7–10
    parameters, 5–7
  special subject, 4, 95
  symbol chart, 98
  thematic, 2, 4, 5
  topographic, 2, 3, 83–111
  USGS index, 36
Mappable features, 95–98
Mapping, field, 57–63
MARC, 23
Marginal map information, 103–8, 113
McDermott, Paul D., 2, 7
Metric coordinate reader, 92–93
Models, terrain, 2
Moffat, Riley M., 29
Muehrcke, Phillip, 5
Mullan Road, 7, 9

Nadir, 15
National Archives, 7
National Cartographic Data Base, 114
National Cartographic Information Center (NCIC), 7, 10
National coordinate systems, 6
National geodetic vertical datum, 105
National map sources, 7–10

National Union Catalog, 7, 23
Nebenzahl, Kenneth, 29
Newberry Library, 31
Nominal scene area, 19, 52
Nominal scene center, 19, 52
North
  magnetic, 59
  true, 59
Northing, 90

Obtaining aerial photographs, 11–12
Ordering LANDSAT-SPOT imagery, 19
Orthophotoquad, 108–9

Panoramic maps, 23
Partial revision, 109
Path
  identifier, 15
  numbers, 19
Phillips, Philip Lee, 23
Photogrammetrists, 114
Photographic coverage request form, USGS, 18
Photoinspection criteria, 109–11
Photorevision, 109
Physical scientist, 114
Plane-table maps, 62–63, 79–82
Principal Meridian, 85–86
Profile, 99, 102
  base line, 103
  cross-sectional, 121–23
  sheet, 123
Projection and grid labels, 106–7
Published maps, 5
Publishing agency, 107–8

Quadrangle, 3, 10
  comparative coverage, 84
  location map, 105
  name, 103
  orthophoto, 108–9
  system, 84–85

Rabenhorst, Thomas D., 31
Range and township numbers, 108
Reconstruction
  foot-made maps, 71
  compass traverse, 73
Reference maps, 3
Relief information, 98–103
Representative fraction, 6
Resource evaluation, 7
Revision
  map, 108–11, 127–31
  partial, 109
  photo, 109
  total, 108–9
Revolutionary War maps, 29
Road symbols, 105
Roamer, 92
Row numbers, 19

Satellite imagery, 7, 10, 15–19
   acquisition, 18–19
Scale, 6, 83, 104
   formats, 6
   horizontal, 103
   vertical, 103
Secondary sources, 22
Sections, 85–86
Sensors and spectral bands, 15
Series, topographic map, 97
Sherman, John C., 2
Sight rule, 62
Simonetti, Martha L., 29
Sketch, field, 58
Small-scale maps, 83
Sohon, Gustuvus, 9
Soundings, 105
Source maps, 5
   parameters, 5–7
Source materials, 2, 6–7
Sources of maps and remote sensing imagery, 32–34
Special-subject (purpose) map, 3, 4, 10, 95
SPOT, 10, 12, 18, 20–21
SPOT grid reference system, 20
SPOT, ordering, 19
State Plane Coordinate System (SPCS), 6, 92–96
   zone-grid, 106–7
Statistical data, 7
Stephenson, Richard W., 22, 23
Subject material, 7
Sun angle, 15
Survey date criteria, 111
Symbols, road, 105

Terrain base, 7
Terrain models, 2
Thematic maps, 2, 3, 4, 5
Thompson, Morris M., 112

Title block, 104
Topographic maps, 2, 3, 10, 83–111, 112
   data, 7
   elements, 113
   interpretation, 117–19
   mapping operations, 113–14
   series, 97
Total revision, 108–9
Township-range blocks, 85
Traverse, 58
   closed, 58–59
   open, 59, 61
Triangulation method, 61
True north, 59

United States Geological Survey (USGS), 3
United States Public Land Survey (USPLS), 85–87
Universal Transverse Mercator (UTM), 6, 86, 88–92
Universal Transverse Mercator Zone, 89
US Bureau of Census, 31
USGS, 10
   index maps, 12–13, 36
   map reference code, 14
   map series coverage, 4
   photographic coverage request form, 18
UTM grid, 107

Vertical datum, 105
Vertical exaggeration, 103
Vertical scale, 103

Wheat, James C., 29
Wolter, John A., 22
World Data Base I & II, 115
Worldwide Reference System (WRS), 19, 21, 51–52
WRS index sheets, 51

Zone grid, state plane coordinate system, 106–7